Efficient Cognition

Efficient Cognition

The Evolution of Representational Decision Making

Armin W. Schulz

The MIT Press
Cambridge, Massachusetts
London, England

This book was set in Stone Serif by Westchester Publishing Services. Printed and bound in the United States of America.

Library of Congress Cataloging-in-Publication Data
Names: Schulz, Armin W., author.
Title: Efficient cognition : the evolution of representational decision
    making / Armin W. Schulz.
Description: Cambridge, MA : MIT Press, [2018] | Includes bibliographical
    references and index.
Identifiers: LCCN 2017029327 | ISBN 9780262037600 (hardcover : alk. paper)
Subjects: LCSH: Decision making. | Human behavior. | Behavioral assessment.
Classification: LCC BF448 .S39 2018 | DDC 153.8/3--dc23 LC record available
    at https://lccn.loc.gov/2017029327

10 9 8 7 6 5 4 3 2 1

For Kelly, my partner in everything

Contents

Acknowledgments

There are a number of people that deserve thanks in bringing this book together. In particular, Justin Garson, Sarah Robins, Corey Maley, Shannon Spaulding, and Richard Bradley read large portions—or even all—of the manuscript and provided insightful and detailed comments. On top of this, Elliott Sober, Dan Hausman, Gualtiero Piccinini, and audiences at the University of Missouri–St. Louis, Kansas State University, the University of Missouri–Columbia, Oxford University, and the 2016 Philosophy of Science Association meeting discussed many aspects of the book with me. There is no doubt that the insights and criticisms of all of these people have resulted in a much improved manuscript. It is a great joy and fortune to have colleagues like these to draw on for help and clarification, and I appreciate all of their support with this project. The same goes for the referees for MIT Press, who have made many very helpful and constructive suggestions that have led to a significantly better book. Philip Laughlin was a fantastic editor to work with. Finally, I want to acknowledge, with thanks, the financial support I have received from the University of Kansas in the form of University of Kansas General Research Fund allocation No. 2302029, which allowed me to spend a summer writing the core parts of this book.

Parts of this book draw on some of my previous work. In particular, parts of chapters 4, 5, and 6 draw on my published articles "The adaptive importance of cognitive efficiency: An alternative theory of why we have beliefs and desires" (*Biology & Philosophy*, 2011, 26: 31–50) and "The benefits of rule following: A new account of the evolution of desires" (*Studies in History and Philosophy of Science Part C: Studies in History and Philosophy of Biological and Biomedical Sciences*, 2013, 44: 596–603). Chapter 8 draws somewhat on my "Gigerenzer's evolutionary arguments against rational choice theory: An assessment" (*Philosophy of Science*, 2011, 78: 1272–1282). Chapter 9 draws on my "Altruism, egoism, or neither: A cognitive-efficiency-based evolutionary biological perspective on helping behavior" (*Studies in History and Philosophy of Science Part C: Studies in History and Philosophy of Biological and Biomedical Sciences*, 2016, 56: 15–23). In all cases, I have significantly revised and expanded the previous treatments, up to the point that there is generally fairly little remaining from the original papers. (This is thus more of a case of descent with *much* modification.)

Finally, I want to thank James and Elizabeth, who supported the work on this book in their own ways. They are wonderful.

I   Foundations

1 Investigating the Evolution of Representational Decision Making

Here is a description of how a particular human being made a difficult decision:

Charlotte herself was tolerably composed. She had gained her point, and had time to consider of it. Her reflections were in general satisfactory. Mr. Collins, to be sure, was neither sensible nor agreeable; his society was irksome, and his attachment to her must be imaginary. But still he would be her husband. Without thinking highly either of men or of matrimony, marriage had always been her object; it was the only honourable provision for well-educated young women of small fortune, and however uncertain of giving happiness, must be their pleasantest preservative from want. (Austen, *Pride and Prejudice*, chap. 22).

Here is a description of how a particular type of ant made a difficult decision:

When a corpse of a *Pogonomyrmex barbatus* worker is allowed to decompose in the open air for a day or more and is then placed in the nest or outside near the nest entrance, the first sister worker to encounter it ordinarily investigates it briefly by repeated antennal contact, then picks it up and carries it directly away towards the refuse piles. [. . .] Almost any object possessing an otherwise inoffensive odor is treated as a corpse when daubed with oleic acid. This classification extends to living nestmates. (E. O. Wilson, 1971, pp. 278–279)

Compared to the *P. barbatus* ants, what makes the way in which Charlotte decides what to do so different—and so fascinating—is that this way of making decisions is based on *mental representations*. Charlotte's actions are not simply a direct function of her perceptions at the time: she does not just nod her head after she sees Mr. Collins going down on one knee. Rather, she takes a step back from her perceptions and assesses the nature of the world she lives in, what life as Mr. Collins's wife would be like, and what she is to do when faced with a marriage proposal (taking into account the constraints on female action of the particular society she is in). It is this sort of non-perceptual assessment that drives her actions.

The *P. barbatus* cleaner ants make decisions quite differently though: they *do* react to the world through their perceptions only. That is, they do not make their behavior dependent on whether it is in fact the case that there is a dead nestmate in front of them, or whether cleaning up dead nestmates is an appropriate thing to do. Rather, the smell of decaying nestmates (whether actual or synthetically produced) is like a switch that turns on their cleaning behavior: the smell simply *triggers* the cleaning behavior. Their behavior is a function only of their perceptual states, without the mediation of intervening mental states.

Put slightly more precisely: representational decision making—as I shall call Charlotte's way of making decisions—is remarkable, as it employs a very different kind of resource to manage interactions with the environment: higher-level, non-perceptual mental states with content. Representationally driven organisms do not just interact with the world based on how they perceive it, but determine what to do using mental states downstream from these perceptual states.

Representational decision making is not a purely human ability, though: it is now widely accepted to characterize the decision-making mechanisms of a number of different organisms. For example, here is a description of how some dogs made a decision about what to do in a certain situation:

When a group of canines in one study learned that a bowl placed on one side of the room contained a treat and a bowl on the other side contained nothing, some of the dogs just sat there when the empty bowl was placed in the center of the room; they figured it was empty and didn't waste their time. These same dogs evinced what researchers said was a similar pessimistic attitude when their masters left for work: They were more likely to howl and tear up the couch when their owner disappeared, possibly because they didn't believe their master would return. (Grimm, 2014)

While the complexity of the contents of the dogs' thoughts seems quite different from that of the contents of Charlotte's thoughts, it is noteworthy that the dogs are still seen to have higher-level thoughts with content. Put differently, the dogs do not just tear up the couch when faced with certain sights and sounds; instead, they do these things after forming (possibly false) thoughts about when their master is coming back. As will become clearer in the rest of this book, much the same goes for many other animals as well.

Given this, the question asked in this book is: why would organisms rely on mental representations to make decisions about what to do? What evolutionary benefits does this reliance provide to a decision maker? This question gains sharpness from the fact that non-representational decision making—i.e., decision making that uses perceptual cues to trigger a behavioral response, as in the case of the *P. barbatus* ants—is a perfectly successful way of making decisions as well: it is fast, easily accomplished, able to lead to complex behaviors, and known to be widely instantiated in the biological world. So why did some organisms adopt a different way of making decisions as well? What does representational decision making provide to these organisms that non-representational decision making does not? In this book, I suggest answers to these questions.[1]

Specifically, I argue that a key factor driving the evolution of representational decision making is cognitive efficiency. In some contexts, representational decision making allows organisms to (a) adjust more quickly and with fewer costs to a changed environment, and (b) streamline the neural machinery underlying their decision-making systems. Furthermore, I argue that these benefits sometimes—but not always—outweigh the disadvantages of this way of making decisions, such as reduced decision-making speed and increased need for cognitive resources like concentration and attention.

Before laying out this account of the evolution of representational decision making in more detail, though, it is worthwhile to address two preliminary questions. First, there is the question of why investigating the evolution of representational decision making is useful and interesting. Second, there is the question of what kind of project investigating the evolution of representational decision making is. Answering these questions is the goal of the rest of this chapter.

The Importance of Investigations into the Evolution of Representational Decision Making

Why should we care about why representational decision making evolved? I think there are two main answers that can be given to this question.

First, finding out about what affected the evolution of representational decision making is intrinsically interesting and important. In general, we care about the factors that influenced the evolution of traits that are widespread, complex, and very important for how organisms interact with their environment. For example, we just want to know, for its own sake, what influenced the evolution of traits like color vision, the ability to generate poison for attack or defense, or of sperms that are covered in a protective shell (angiosperms). The reason for this is that widespread, complex, and ecologically important traits like these are likely to be adaptively valuable in a number of different environments, but also relatively difficult to evolve due to the fact that they depend on many different parts coming together in the right ways (see also Godfrey-Smith, 2001; Dawkins, 1986).[2]

This matters here, as there is every reason to think that representational decision making is a trait that is like this. First, as I will make clearer in chapter 2 (as well as in chapters 5 and 6), there is good reason to think that representational decision making is a complex trait: it is underwritten by a distinctive cognitive and neural organization with many different elements interacting in many different ways. There is also good reason to think that this trait is ecologically and adaptively very important: after all, decision making—the way an organism generates its behaviors—is a central feature of an organism's life that it cannot avoid, and that can greatly impact its success in interacting with its environment (again, I return to this point in chapters 5 and 6). Finally, as I argue in chapter 2, there are good reasons to think that representational decision making is a trait that is found in many different groups of organisms: from spiders and rats to macaques and humans, many different organisms plausibly at least sometimes make decisions by relying on mental representations. In short, representational decision making is a trait whose evolution is inherently interesting and worth investigating.

However, the relevance and importance of the investigation into the evolution of representational decision making goes beyond its intrinsic interest. This is due to the fact that this investigation is also instrumentally valuable: it can be helpful in getting clearer on a number of other issues as well. To see this, note that while there is much about representational decision making that is quite well understood, there are also a number of questions concerning this way of making decisions that are still much debated. Consider three of these open questions in a bit more detail (for a discussion of various further questions surrounding representational decision making, see, e.g., Nichols & Stich, 2003; Allen & Bekoff, 1997; Schroeder, 2004; Prinz, 2002).

First, much recent work in cognitive science, psychology, biology, and philosophy has emphasized the idea that, in order to understand the way an organism thinks, it is important to see this organism as situated within its environment. Put differently: cognition has increasingly come to be seen as "extended"—as so enmeshed with features of the organism's environment that it is partially constituted by these external features (see, e.g., Clark, 1997, 2008; Rowlands, 2010; Shapiro, 2004; Griffiths & Stotz, 2000; Stotz, 2010; Schulz, 2013; but see also Adams & Aizawa, 2008; Rupert, 2009). There are many different versions of this kind of view (Menary, 2010; R. Wilson, 2004; Rupert, 2009), but for present purposes, it is sufficient to note two general aspects of it. On the one hand, there is a significant strand within this body of work suggesting that extended cognition and representational cognition are in direct opposition to each other: to the extent that an organism's decision-making processes are constituted by environmental features, to that extent they are not representational—and vice versa (Brooks, 1991; van Gelder, 1995). On the other hand, a number of authors in this area argue that much of cognition should be seen to be extended (and thus non-representational): according to these authors, non-extended, representational decision making should be taken to be a biological oddity that has rarely evolved (see, e.g., Brooks, 1991; Chemero, 2009).

Now, it turns out that knowing about (some of) the reasons for why representational decision making has evolved is very useful for making the issues surrounding extended decision making clearer as well. In particular, knowing about the factors influencing the evolution of representational decision making can help make clearer that it is not true that we should expect representational decision making to be a biological oddity. In fact, representational decision making requires the same *sorts* of explanations as non-representational decision making. Furthermore, understanding the factors driving the evolution of representational decision making also helps make clearer that representational and externalized decision making should not be seen to be in opposition to each other: in fact, there are circumstances in which it is plausible to expect organisms to rely on decision-making mechanisms that are both representational and that rely heavily on the organisms' environments. Making this clearer is the aim of chapter 7.

The second open question surrounding representational decision making relevant here concerns whether human and non-human animals should be seen to rely on a handful—or even just one—highly general (and typically optimizing) decision rule, or whether they should be seen to rely on very many highly specialized (and typically satisficing) decision rules (see, e.g.,

Gigerenzer & Selten, 2001; Stanovich, 2004; Tversky & Kahneman, 1974; Chater, 2012; Kacelnik, 2012). That is, when determining whether to accept a potential mate, do organisms rely on the same (perhaps optimizing) decision principle that they rely on when deciding whether to leave their current foraging patch for a new one, or are these two decisions based on two different (perhaps "satisficing") rules? These issues are currently much discussed, both concerning humans (see, e.g., Gigerenzer & Selten, 2001; Stanovich, 2004) and non-human animals (see, e.g., Kacelnik, 2012).

Again, knowing something about the evolutionary pressures on representational decision making can help advance this discussion. This is because knowing about when and why representational decision making is likely to be adaptive might also tell one something about which kind of representational decision making—general or specialized—is likely to be adaptive in which circumstances. More specifically, as I make clearer in chapter 8, there is reason to think that the situations particularly suited to generalist decision making concern choices about how to live in a complex social group, whereas the situations particularly suited to specialized decision making concern choices about how to interact with various tools and about which resources to consume in what manner. In turn, this last set of insights can be used both as the basis of further studies and as evidence for existing hypotheses concerning the generalist or specialist nature of human and animal decision making.

Finally, there is a longstanding set of disputes concerning the psychology of helping behavior, both in humans and in other animals. In particular, there is a long tradition in psychology and philosophy considering the question of whether humans (and other organisms) are disposed to help others—where they are so disposed—for egoistic or non-egoistic reasons (see, e.g., Batson, 1991; Stich et al., 2010; Sober & Wilson, 1998). So, when organism A helps organism B, is that because A reasoned that helping B is somehow in its own interest—perhaps because A's helping B now will dispose B to help A in the future, or perhaps because helping B will make A look more attractive to potential mates? Or does A decide to help B simply because A cares about B's well-being—independently of whether helping B will also benefit A in some way? While much work has gone into addressing these questions, the overall upshot here is still unclear (see, e.g., Stich et al., 2010).

However, again, an understanding of the evolutionary pressures on representational decision making can be helpful here (especially because it has turned out to be difficult to get empirical evidence for or against the existence of psychological altruism—see, e.g., Cialdini et al., 1997). This may not be obvious at first, since the dispute concerning the psychology of helping behavior may not appear to be about representational decision making per se. However, as I show in chapter 9, a closer look reveals that representational decision making—or its absence—is in fact at the heart of this dispute. This is because the core difference between egoists and non-egoists is precisely the fact that the former invests more heavily in representational decision making: whereas the non-egoist is simply inherently disposed to help another organism, the egoist relies on higher-level mental representations to determine whether it should help that other organism. With this is in mind, knowing about when and why representational decision making is adaptive can tell one more about when and why being an egoist is adaptive—and when not. In this way, it becomes possible to make some progress in the debate surrounding psychological egoism and non-egoism: in particular, it becomes clearer in which circumstances organisms ought to be expected to be altruistically motivated, in which circumstances they ought to be expected to be egoistically motivated, and in which circumstances they ought to be expected to be motivated in a way that is neither altruistic nor egoistic. Chapter 9 is dedicated to exploring this point in more detail.

All in all, therefore: investigating the evolution of representational decision making is both inherently and instrumentally useful. It is inherently useful, since we simply care about what factors plausibly had an impact on the evolution of complex, widespread, and adaptively important traits like representational decision making. It is instrumentally useful, since knowing something about what has affected the evolution of representational decision making can aid in the investigation of various other open questions surrounding representational decision making—such as the relationship between the extendedness of cognition and its representational nature, the structure of an organism's decision rules (general and optimizing or specialized and satisficing), and the nature of an organism's motivation to help others.

The Nature of Investigations into the Evolution of Representational Decision Making

Before getting started on this investigation, though, a few notes about its nature are in order. Given its content, there can be little doubt that this investigation is, at least to a significant extent, an empirical project:

determining the major pressures on the evolution of representational decision making is not something that can be accomplished just from the
armchair. However, this does not mean that abstract and theoretical considerations are not integral to this project as well.[3]

The main reason for this is that this is not a project that can be
addressed with the methods of just one science. Since representational
decision making is widespread and important to many different aspects
of many different organisms' lives—from foraging and mating to parental
care and social interactions—the investigation of its evolution is something
that needs to draw on many different sciences, from evolutionary biology
and ecology to anthropology, psychology, social science, and philosophy.

In turn, this implies that a large part of the cognitive labor that goes
into this project involves the assessment of how the different considerations
from the different sciences can be integrated. To do this, though, one needs
to draw on relatively abstract and theoretical arguments; it is not just a matter of doing more studies of one kind or another. Put differently: much of
the work involved in laying out an account of the evolution of representational decision making involves *synthesizing* a wide variety of insights into
a coherent whole.

Furthermore, the interdisciplinary nature of the investigation into the
evolution of representational decision also implies that methodological
considerations become very prominent. Since, as just noted, this investigation requires bringing together many different sciences, it becomes important
to consider what role the considerations from one science can play in the
investigation of issues in another science. In particular, in using evolutionary biological considerations to push forward disputes in cognitive science,
social science, and philosophy, we need to be clear about the impact these
considerations can have in these other sciences. This point is especially
important to note since evolutionary psychological projects like the present one are, undoubtedly, controversial: a number of authors have argued
that projects like it *must* be overly speculative and therefore epistemically
unconvincing (see, e.g., Buller, 2005; Richardson, 2007).

However, as I try to show in the rest of this book, I think this conclusion
is overly negative. While it may be true that highly interdisciplinary, highly
abstract, and highly theoretical projects like the present one cannot avoid
being somewhat speculative—since the issues here are so abstract, we are

often forced to move ahead of the data—this does not mean that such projects therefore necessarily are epistemically or predictively worthless. After all, there is a big difference between pure fantasy and empirically grounded and clearly reasoned (if sometimes somewhat speculative) arguments. Bringing out that it is possible to provide the latter is thus one of the key aims of this book. Of course, I am under no illusion that I will be able to end the controversy surrounding evolutionary psychology once and for all. I hope, though, that, at the very least, my book can reinvigorate the debate surrounding (broadly) evolutionary psychological research by adding a novel, extended perspective on the evolution of representational decision making.

Conclusions

In this book, I present and defend a view of some of the major adaptive pressures that shaped the evolution of representational decision making and apply it to a number of open questions concerning this psychological trait. Before getting started on this—at least in my view—exciting and illuminating project, it is useful to say a few words about how the book is structured.

The book has three parts. Part I (chapters 1–3) lays out the foundations of the account of the evolution of representational decision making defended in the book—specifically, it makes clearer what sort of project investigating the evolution of representational decision making is (in this chapter), what exactly I understand representational decision making to be like (in chapter 2), and why asking about the evolution of this sort of trait is defensible (in chapter 3). In part II (chapters 4–6), I present my account of the evolution of representational decision making (in chapters 5 and 6) and critically discuss some of the existing treatments of this issue already in the literature (in chapter 4). In part III (chapters 7–9), I apply the account of part II to the three open questions concerning the nature of representational decision making mentioned above: the extendedness of decision making (in chapter 7), the specialization of decision making (in chapter 8), and the psychological sources of helping behavior (in chapter 9).

In reading the book, it is useful to realize that the three parts support each other. Most obviously, part I ensures that the account of part II stands on a solid foundation, and part III applies the insights of part II to a number of open questions in the literature. However, it is also the case that the account of part II can be seen to spell out some of the more abstract remarks of part I, and that part III can be seen to further clarify several aspects of the account in part II. For this reason, rather than seeing the book as a collection of independent chapters, I think it is better to see the book as a unified whole.

With this in mind, it now becomes possible to consider more closely exactly what I understand by representational decision making. Doing this is the aim of the next chapter.

2 The Nature and Reality of Representational Decision Making

Some organisms interact with their environment by relying on a battery of stored, relatively unmediated behavioral responses to various environmental contingencies. A particular feature of the environment—a certain smell, say—is detected by a dedicated sensor of the organism; the state of this sensor then triggers a biochemical cascade leading to a motor command in the organism's motor cortex; and the motor command results in a particular form of behavior (Tulving, 1985; Lieberman, 2003; Graybiel, 2008; Shettleworth, 2009; Dickinson & Balleine, 2000).

Other organisms—or the same organisms in other circumstances—do something else, though: they rely on mental content.[1] Specifically, instead of just reacting to the states of their sensory or internal monitoring systems, these organisms make their behavioral responses dependent on higher-level mental states. They use their sensory or internal monitoring systems merely as *inputs* into the determination of how they should react to the world; this determination itself, though, is based on distinct states that are downstream from their perceptual systems (Allen & Bekoff, 1997; Allen & Hauser, 1991; Carruthers, 2006; Tulving, 1985; Sterelny, 2003; Rosenzweig, 1996; Millikan, 2002).

It is the goal of this chapter to make the nature and reality of these two ways of making decisions clearer. More specifically, I first specify how representational decision making is to be distinguished from non-representational decision making. Given this, I then lay out why I think that both of these ways of interacting with the environment are biologically and psychologically real traits whose evolution is worth discussing.

The chapter is structured as follows. In the first section below, I make clearer how the difference between representational and non-representational decision making is to be understood. Next, I make "reflexive," non-representational decision making more precise. Then, I make representational decision making more precise. In the following section, I support the reality of representational decision making by appeal to a wide range of literatures from a number of different disciplines. Finally, I use the insights of the previous two sections to bring out several widely accepted features of representational decision making. I summarize this discussion in the final section.

Representational versus Non-representational Decision Making

In this book, I use a relatively common architectural criterion to mark the difference between representational and non-representational decision making. More specifically, in what follows, I take it that:

an organism is a *non-representational decision maker* to the extent that its behavior is an immediate function of its perceptual states; and

an organism is a *representational decision maker* to the extent that its behavior is an immediate function of various higher-level states downstream from its perceptual states.

As it is understood here, therefore, the distinction between representational and non-representational decision making distinguishes behavior that is simply *triggered* by the perception of the state of the world from behavior that is derived from mental states that are downstream (and thus distinct) from these perceptual states. A non-representational decision maker reacts to the world directly (or as directly as its perceptual systems will allow it), whereas a representational decision maker reacts directly only to its intermediate, higher-level mental states. So, for example, an organism for which the smell of blood directly triggers attacking behavior is a non-representational decision maker, whereas an organism for which the smell of blood merely triggers an intermediate state (perhaps one representing the fact that there is wounded prey nearby), which then becomes the driver of the organism's behavior, is a representational decision maker.

As noted earlier, this way of marking the distinction between representational and non-representational decision making is relatively common in the literature. So, for example, Allen and Hauser (1991, p. 231) state:

Humans are capable of recognizing something as dead because they have an internal representation of death that is distinct from the perceptual information that is used as evidence for death. It is this separate representation that is capable of explaining the human ability to reason about death rather than merely respond to death in the environment. . . . We would attribute an abstract concept to an organism if there is evidence supporting the presence of a mental representation that is independent of solely perceptual information.

Related ideas have been expressed by a number of other authors as well (see, e.g., Prinz, 2002, pp. 256–257; Tulving, 1985; Allen, 1999; Dickinson & Balleine, 2000; Gallistel & King, 2009; Penn et al., 2008; Whiten, 2013; Martinez-Martinique, 2014). For this reason, this way of drawing the distinction between representational and non-representational decision making should not be seen as highly idiosyncratic or outlandish. Still, it is useful to make a few more points about this distinction here.

First, the above characterization of representational and non-representational decision making depends on there being a relatively robust distinction between perceptual states and non-perceptual states. This, though, is plausible: while recent work in cognitive neuroscience has made it clear that perceptual processing is complex and hierarchical already (see, e.g., Felleman & van Essen, 1991; Marr, 1982), this work generally still maintains a distinction between purely perceptual processing and non-perceptual processing (see, e.g., Borensztajn et al., 2014; O'Reilly et al., 2014; Pylyshyn, 1999; Firestone & Scholl, 2016). In other words, while it has become clear that a perceptual state needs to be seen as involving spatially and temporally extended layers of neural and cognitive processes, this does not mean that no distinction between a perceptual state and a higher-level state that is downstream from this perceptual state can be made. (The sections below return to this point.) Nothing in what follows depends on the details of how the distinction between perceptual and non-perceptual states is to be drawn; all that matters here is that some such distinction is accepted. (In fact, it is possible to rephrase the core ideas of this book in terms of merely a distinction in degree between perceptual and non-perceptual states—this would then make the difference between representational and non-representational decision making one of degree only as well. While adding some complexities, all conclusions derived in the rest of the book would remain the same.) However, that said, it is important to note that I here wish to characterize "perceptual states" broadly enough to cover internal states as well (i.e., what might be called "apperception"). Put differently, I assume that non-representational organisms can monitor ("perceive") both their external and their internal environment, and react to the outcomes of this monitoring process directly.

The second point to note concerning the distinction between representational and non-representational decision making is that I leave open exactly what grounds the content of the relevant inner states. That is, I here do not commit myself to an account of what makes it the case that a given mental state is a mental representation of p. So, for all that follows, a mental representation can be grounded in any of the following (see, e.g., Crane, 2016).[2]

(a) Teleosemantic accounts (Millikan, 1984; Papineau, 1987; Neander, 2006): higher-level state S in organism O has content C if the tokening of S when the world is in state C has been selected for in the population to which O belongs.[3]

(b) Causal accounts (Stampe, 1986; Dretske, 1981; Rupert, 1999): higher-level state S in organism O has content C if the world's being in state C causes, in appropriate conditions, the tokening of S in O.

(c) Law-based accounts (Fodor, 1990): higher-level state S in organism O has content C if the tokening of S is asymmetrically dependent on non-C-caused tokenings of S (i.e., the latter tokenings would not happen if C-caused tokenings did not exist).

(d) Developmental accounts (Dretske, 1988): higher-level state S in organism O has content C if O has learned to token S when the world is in state C.

(e) Structural accounts (Prinz, 2002; Barsalou, 1999): higher-level state S in organism O has content C if S has the appropriate structure to allow O to detect the world's being in state C.

Nothing in what follows depends on exactly what explains that a mental state has content: I just take for granted that some such explanation can be given, and that this explanation is adequate for the empirical work surrounding the literature on representational and non-representational decision making sketched in the sections below.

Third, however, it does need to be accepted that the present account diverges from the above accounts of the nature of mental representation in requiring the presence of a higher-level mental state. So, on the above accounts of mental representations, purely perceptual states can also be representational: for example, on teleosemantic accounts, any mental state— whether higher-level or perceptual—can be representational, as long as it has the right evolutionary history (Millikan, 1984, 1989, 1990). However, I am happy to accept this consequence, and restrict my inquiry to the evolution of decision making based on higher-level mental representations— that is, on mental representations that are downstream from the organism's perceptual states. My goal is an account of the evolutionary pressures on certain ways minds can be *structured*; it is not an investigation into the (supposed) evolutionary nature of mental content or, more generally, into how the existence of meaning or content can be squared with a purely physical universe. For this reason, the restriction to higher-level mental representations is unproblematic—it is just part of the specification of the project I am engaged in. The question I am trying to get closer to answering is why some organisms sometimes rely on mental states downstream from perceptual states, and not just on the perceptual states themselves. This is an interesting question independently of whether the latter are—or are not—to be seen as representational.[4]

Relatedly, it does need to be acknowledged that my account does not guarantee that the higher-level states that drive the behavior of a representational decision maker have the features needed to qualify as mental representations on all of (a)–(f) above.[5] In particular, my account would classify "swamp creatures"—organisms without any learning or evolutionary history—as representational decision makers as long as they in fact acted on higher-level mental states downstream from their perceptual states. However, since my project is the investigation of the evolution of a certain kind of mental architecture, this issue, too, is not greatly problematic (see also Millikan, 1996; Neander, 1996). While I think that the architecture at the focus of my inquiry is well described as being representational in nature— as noted above, this fits to how much of the rest of the literature carves up the issues—the labeling of this architecture is, at the end of the day, not greatly important. (If so desired, representational decision making could also be called "non-perceptual" decision making or "higher-level" decision making.) So, while I shall continue to refer to "mental representations" when setting out my theory, unless further specified, all such references can just be seen to refer to higher-level mental states that mediate between perceptual states and behavioral responses, and could be replaced with the latter description if so desired (see also Camp, 2009).

In all: representational decision makers, as understood here, differ from non-representational decision makers in reacting not to their perceptual states directly, but to higher-level states that mediate between their perceptual states and their behavioral responses.[6] With the difference between representational and non-representational decision makers thus clarified, consider now these two types of decision makers in slightly more detail.

Non-representational Decision Making

Not every organism is a representational decision maker—and certainly not all the time. In fact, probably all organisms are non-representational decision makers some of the time. Given this—and also so as to come to a better understanding of the nature of representational decision making—it is important to make the nature of non-representational decision making more precise.

As noted in the previous section, non-representationally driven organisms make decisions by relying on stored mappings between (broadly understood) perceptual states and specific behavioral outcomes (Lieberman, 2003; Graybiel, 2008; Carruthers, 2006; Shettleworth, 2009; Godfrey-Smith, 1996; Alcock, 2013; Heyes, 2013). Put differently, non-representational decision makers let their behavioral response to an environmental feature simply be triggered by the non-representational detection of that feature. Graphically, this can be represented by the following "table of reflexes" (I return below to the choice of this terminology):

Note that the way of making decisions depicted in table 2.1 is non-representational in two senses: not only is the trigger of the organism's behavior not based on a higher-level mental state (it is just based on a perceptual state), but the process of triggering itself is also not based on such a

Table 2.1
A Table of Reflexes

Perceptual Cue	Behavior
Visual pattern 1	Behavior 1 (e.g., freeze in place)
Visual pattern 2	Behavior 2 (e.g., fly away)
Tactile pattern 1	Behavior 3 (e.g., initiate mating behavior)
Tactile pattern 2	Behavior 2
Tactile pattern 3	Behavior 1

higher-level state (it is just based on a mapping). This is important to note, as it implies that an organism can be a representational decision maker in three different ways: (a) it could let the tokening of certain *mental representations*—not perceptual states—trigger a behavioral response, but this triggering could still be non-representational; (b) it could rely on a mental representation to mediate between non-representational perceptual states and behaviors; and (c) it could rely on both (a) and (b). I return to these issues below and in chapters 4–6, but for now, I concentrate on purely non-representational decision making—that is, decision making that, as in table 2.1, results from a mapping between an organism's perceptual states and particular forms of behavior. Five points are important to note about this way of interacting with the environment.

First, there can be no doubt that this is a widespread way of generating behavior: many types of organism are well seen to make at least some decisions purely non-representationally. This can be seen from the large and variegated list of organisms some of whose behaviors have been successfully accounted for in purely non-representational terms. The following examples illustrate the many different types of organisms and many different forms of behavior that have been illuminated in this way.[7]

1. Many bacteria are known to interact with their environment by relying on triggered responses to perceptual states. For a famous example, consider magnetotactic marine bacteria, which determine where to swim to by letting the state of their magnetosomes trigger the relevant swimming behavior (Simmons et al., 2006; Faivre & Schüler, 2008; Blakemore, 1975).[8]

2. Some microbes make decisions by "quorum sensing": the microbes detect the quantity of a particular chemical compound around them, and, if it is below or above a certain threshold, engage in a particular form of behavior (Miller & Bassler, 2001). For example, slime molds can determine whether to aggregate into a "fruiting body" by assessing whether the concentration of the bacteria they eat has fallen below a threshold level; if it has, they begin the process of forming a fruiting body that transports them to a new location (Strassmann et al., 2011; Kuzdzal-Fick et al., 2011).

3. As also noted in chapter 1, worker ants of the species *Pogonomyrmex barbatus* let the presence of oleic acid—which generally correlates in their environment with the presence of decomposing dead bodies—trigger their cleaning behavior: anything that smells of oleic acid (whether alive or dead)

is moved to the refuse pile (E. O. Wilson, 1971, pp. 278–279). Also, many social insects can achieve complex logistical feats—for example, building complex structures—by letting behavioral responses be triggered by the pheromonal traces of their conspecifics (see, e.g., Theraulaz & Bonabeau, 1999; Buhl et al., 2005; Garnier et al., 2007).

4. In many birds, feeding behaviors seem to be triggered by the sight of begging chicks (I. G. Jamieson & Craig, 1987; Koenig & Mumme, 1990).

5. Predatory behavior in many (though not all) amphibians and reptiles is triggered by the presence of the appropriate perceptual cues: for example, tongue protrusion in various salamanders and frogs is triggered by the presence of the appropriate visual stimuli (Deban et al., 2001; Lauder & Reilly, 1994; C. W. Anderson, 1993).

6. Many conditioned or associatively learned behaviors in many mammals are non-representationally driven: for example, rats can learn to associate a given sound with the presence of food, so that the presence of the sound simply triggers foraging behavior (Dickinson, 1985; Dickinson & Balleine, 2000; Dickinson et al., 1995; see also McEchron et al., 1998; Levy & Steward, 1979). Also, many primates seem to make many decisions purely by association learning: a particular sight is non-representationally associated with a given behavioral response (Byrne, 2003).

7. Many human actions are non-representationally driven. For example, skilled sports players react to environmental conditions by letting certain stimuli trigger behavioral responses (Abernethy & Russell, 1987; Fischman & Schneider, 1985; A. M. Williams & Hodges, 2004).

There is no doubt that much more could be said about each of the examples above. However, for present purposes, the key point to note is that they show there is much good theorizing that can be done by seeing some organisms as being non-representational decision makers (at least sometimes): many organisms' behaviors are well explained and predicted by seeing them as triggered by the non-representational detection of some environmental feature.[9]

The second point to note concerning non-representational decision making is that it is quite powerful: in particular, it can lead to highly complex behaviors. For example, consider the complex physical constructions accomplished by many social insects, the complex coordination achieved by slime molds, and the complex conditioned responses displayed in rats: these behaviors can be temporally and spatially quite extended, involve interactions with others, and require a number of different behavioral steps to be completed. Still, many of these behaviors—most obviously ones that are associatively learned—can be acquired very quickly and easily: indeed, it is known that many animals can be taught very quickly to engage in specific behavioral routines when faced with specific, purely perceptual stimuli such as sounds, sights, or smells (Mackintosh, 1994; Dickinson & Balleine, 2000).[10]

Third, and relatedly, it is important to note that it is not the case that, in non-representationally driven organisms, the perception of a given state of the environment always triggers the same behavioral response. A major reason for this lies in the fact that the organism's "internal" state also matters for which behavioral response is triggered: an organism that is hungry might react differently to the smell of food than one that is not hungry. There are two ways in which this can be understood: either the organism relies on a mapping between *combinations* of perceptual states and internal states (e.g., smells *and* hunger) and behaviors to make decisions (see also Papineau, 2003), or the mapping between perceptual states and behaviors that the organism relies on is dependent on its internal states (so that the presence of hunger switches out the mapping between smells and behaviors that is in place when the organism is not hungry). For present purposes, it is not necessary to further consider this issue; all that matters here is that it is kept in mind that non-representationally driven organisms can make their behavior dependent on their internal condition as well as their perception of the state of the external environment.

The fourth point to note concerning non-representational decision making is that the neurobiology of this way of making decisions is relatively well understood. In particular, it is known that, at least in mammals, non-representational decision making is especially related to activation in the basal ganglia, the amygdala, the hippocampus, the midbrain, and the striatum (McEchron et al., 1998; Levy & Steward, 1979; Faure et al., 2005; Kim & Hikosaka, 2015; O'Doherty et al., 2004; Casey et al., 2002; O'Reilly et al., 2014; Alcock, 2013, chap. 4; Scott, 2005, chap. 2).[11] So, for example, O'Reilly et al. (2014, p. 201) note that "this kind of network [the basal ganglia, cerebellum, and hippocampus] is fantastic for rapidly processing specific information, dealing with known situations, and quickly channeling things down well-greased pathways."

This is important to keep in mind, in addition, because it provides further support to the fact that non-representational decision making is a genuinely distinctive way of making decisions: the issues here do not just concern appropriate ways of describing a given behavior—they are not just matters of a "physicalist stance" (Dennett, 1987)—and they are not merely the outcome of a commitment to an outmoded form of behaviorism (D. Jamieson & Bekoff, 1992). Instead, they should be seen to have inherent psychological and neurological reality: non-representational decision-making systems are distinctive types of cognitive systems that recruit—at least in mammals—a distinctive set of neural regions.

The fifth and last point that it is important to make here is that several researchers make finer distinctions among non-representationally driven behaviors. For example, several authors distinguish "fixed-action patterns" (stereotyped behavioral sequences) from—among others—"instincts" (innate behavioral dispositions), "reflexes" (direct responses by the nervous system), and "habits" (learned behavioral dispositions; Barrows, 2011; Auletta, 2011; Mayr, 1974; Ristau, 1996). However, for present purposes, these finer distinctions are not so important: nothing in what follows turns on differences among non-representational ways of making decisions; all that matters here are differences among representational and non-representational ways of making decisions. For this reason, while, for expository purposes, I may sometimes write about "reflexive behavior" or "reflexes," this should always be read as referring to non-representationally driven behavior in general, and not to any of its particular variants.

All in all, therefore: non-representationally driven organisms make decisions by relying on a mapping between various perceptual states and specific behavioral responses. This is a powerful and widespread way of making decisions that can be relatively quickly acquired and lead to flexible and complex behaviors, and which is, at least in mammals, underwritten by activation in a specific set of neural regions. With this in mind, consider now the nature of the alternative, representational way of making decisions.

Cognitive, Conative, and Full Representational Decision Making

It is widely accepted that full representational decision making has two aspects: a *cognitive* (indicative) and a *conative* (directive) component (see, e.g., Nichols & Stich, 2003; Hausman, 2012; Chaiken & Trope, 1999; Schroeder, 2004; Sterelny, 2003; Glimcher et al., 2005; Anderson, 2007). However, beyond this basic claim, there is much disagreement about what representational decision making comprises.

First, some writers see the cognitive and the conative aspects of representational decision making as closely intertwined—up to the point of being defined in terms of each other (M. Smith, 1987; Anscombe, 2000; Davidson, 1980). In contrast, others see the two as strongly dissociable (Sterelny, 2003; Millikan, 2002; Schroeder et al., 2010). Now, for reasons that will be made clearer below and in chapters 4–6, it is the latter, separability-based perspective that I adopt here. This is so for two reasons: on the one hand, as will also be made clearer below, the separability of cognitive and conative representational decision making has much neuroscientific and psychological plausibility (Schroeder et al., 2010; Sterelny, 2003; but see also Colombo, 2017); on the other, seeing cognitive and conative representational decision making as separable allows for a plausible evolutionary biological treatment of representational decision making. Since bringing out this second point is one of the core aims of this book, though, it is best to see the separability of cognitive and conative representational decision making as a working assumption for now, whose reasonableness will be supported as we go along.

Second, a number of authors propose to make a number of finer distinctions among the cognitive and conative components of the representational decision-making system (Bratman, 1987; Hausman, 2012; Davidson, 1980). For example, they propose to distinguish among desires, preferences, and values (Hausman, 2012), or among beliefs, aliefs, and imaginations (Gendler, 2008; Sinhababu, 2013). However, since nothing I say in this book depends on such a finer division, I will not consider this further here.

Third, a number of authors argue for the existence of representational states that are not well classified as being either cognitive or conative. For example, Millikan (2002) argues for the existence of pushmi-pullyu representations that are both cognitive and conative (see also Shea, 2014; Zangwill, 2008; and chapter 4 below). Also, Bradley and List (2009) argue for the existence of cognitive representations that have some of the features of conative representations (e.g., concerning how they are updated). While I do not consider these other representational states further here, nothing in what follows conflicts with arguments positing their existence (though see chapter 4 for more on Millikan's view). The one commitment that I do make is that the cognitive and conative representational decision-making systems make for interesting traits to consider on their own as well as together: I assume that the evolution of the cognitive and conative representational decision-making systems can be usefully studied separately from the evolution of other representational decision-making systems. However, given how I understand the former two systems—which will be made clearer below and in chapters 4–6—this should not be greatly controversial.

With this is in mind, consider the cognitive and conative representational decision-making systems separately first, and then conjointly. Note that the goal in this is to make precise in which way I understand these systems, out of the many such ways proposed. While this invariably requires taking a stand as to which ways of understanding these systems are most useful, the goal in what follows is not to come up with some novel characterization of these systems that has been overlooked in the literature. Rather, the goal is to come up with a characterization of these systems that does justice to the empirical literature on representational decision making—aspects of which I sketch below—and which is helpful for the rest of the discussion in the book.

Cognitive Representational Decision Making

Cognitive representational decision making, as I understand it here, concerns decision making that is based on representations about the state of the world. That is, cognitive representational decision makers interact with their environment by considering representations about what state the world is in (for similar understandings, see, e.g., Fodor, 1990; Dretske, 1988; Millikan, 2002; Cummins, 1996; Nichols & Stich, 2003; Papineau, 2003; Burge, 2010; Sterelny, 2003; Allen & Bekoff, 1997).[12] Three points are important to note about this way of understanding cognitive representational decision making.

First, "state of the world" should be understood broadly to not just include the state of the organism's external environment, but also the state of its body—its internal environment. That is, an organism that engages in a particular form of behavior—drinking, say—because it represents its body to be in a certain state—dehydrated, say—is a cognitive representational decision maker. (Note also that, due to the fact that the trigger of the drinking here is a cognitive representation, this is very different from drinking that is triggered by the mere fact that the organism detects that it is dehydrated—that is, without the intervention of a mental representation; see also chapter 5.)

Second, there are no restrictions on the content of the representations driving the behavior for them to qualify as being a cognitive representation.

An organism that makes its behavior dependent on whether it represents its environment as containing "loud, brownish things" is as much a cognitive representational decision maker as one that makes its behavior dependent on whether it represents its environment as containing howler monkeys of the species *Alouatta seniculus* (Dupre, 1996; Allen, 1992). Put differently: being a cognitive representational decision maker is not a matter of relying on representations with any particular kind of conceptual content—it is a matter of relying on mental states that are distinct from the perceptual states of the organism, and which sort these perceptual states in some way (O'Reilly et al., 2014, p. 201; Martinez-Martinique, 2014; Camp, 2009).

Third, it is important to note that it is non-trivial to behaviorally determine whether an organism is a cognitive representational decision maker (see also Sterelny, 2003). In particular, the mere fact that an organism reacts in the same way to many different states of the world does not necessarily mean that the organism is a cognitive representational decision maker. This is so for two reasons. On the one hand, the organism might merely track the environment "robustly" (Sterelny, 2003): several different perceptual states might lead to the same behavioral response. On the other hand, the different environmental states might not lead to genuinely different perceptual states: as far as the organism is concerned, these different states of the world do not look (or sound or smell) different. (This is related to a phenomenon known as "stimulus generalization": see, e.g., D'Amato & van Sant, 1988.) However, this does not mean that it needs to be impossible to determine whether an organism is a cognitive representational decision maker. It just means that what is needed is a relatively large amount of both behavioral and neural-psychological data. (Also see the next section for more on this point.)

In short: cognitive representational decision making concerns the determination of behavior through the consultation of representations about the state of the world, broadly understood. Consider next conative representational decision making.

Conative Representational Decision Making

Conative representational decision making, as I understand it here, concerns decision making that is based on representations about what the organism is to do. Unlike cognitive representational decision makers, then, conative representational decision makers do not (need to) make their behavior dependent on representations about what the state of the world is, but on representations about what their goals are—what they aim to accomplish (for similar understandings, see, e.g., Millikan, 2002; Berkman & Lieberman, 2009; W. J. Smith, 1990; Klaes et al., 2011; Cheng & Holyoak, 1985; Ramnerö & Törneke, 2015; Sterelny, 2003; Schroeder, 2004). There are five points that need to be noted about this way of making decisions.

First, conative representations are "directive" or "imperative" in structure—unlike cognitive representations, which are "descriptive" or "indicative" in structure (Millikan, 2002, pp. 258–259; Prinz, 2002). However, this does not mean that they need to take the form of commands or rules; they could also be based on functions that the organism is computing. The point is just that they represent what is to be done—they do not describe what the world is like, but state what the organism is to do.

Second, just as with cognitive representations, I here leave it open what the content of conative representations is like. In fact, I do not even assume that conative representations need to operate on cognitive representations (also see chapter 6). So, an organism that detects colors in a non-cognitive representational way—for example, by just relying on a non-representational wavelength detector—can still be a conative representational decision maker, in that it could make its behavior dependent on a rule like "if things look like this [i.e., light is reflected with a wavelength of between 400 and 550 nanometers] do x, if things look like that [i.e., light is reflected with a wavelength of between 550 and 700 nanometers], do y." Note that, for this organism to be a conative representational decision maker, the rule itself has to be represented in its cognitive system; an organism that simply does x when perceiving light with a wavelength of between 400 and 550 nanometers is not a conative representational decision maker; for the latter, it would need to do x *because* it considered the above rule.

Third, there are good empirical and theoretical reasons to think that conative representations will be associated with various other mental states—such as certain emotions or rewards (Schroeder, 2004; Schroeder et al., 2010; Morillo, 1990; Damasio, 1994). The reason why these connections are to be expected is that they make for a good explanation of why these kinds of representations are directive in structure: the link to emotions or rewards can provide the "urge towards action" that marks a mental representation as conative (Damasio, 1994; Millikan, 2002; Schroeder, 2004). However, since for present purposes the details of these connections are not so important (with some exceptions concerning the neuroscience of representational decision making noted below), I will not consider them further here.

Fourth, it is important to note that, since the behavior of conative representational decision makers is driven by representations of the goals of the organism, these goals need to be of such a form that the organism can, in fact, act on them. Exactly what this entails depends on the details of the rest of the organism's cognitive system. So, for most humans, the goal of "buying low, selling high" will not be useful in driving investment behavior, since we are generally unable to tell when an investment is undervalued. However, it is conceivable that some humans—for example, ones with information about future financial developments—could act toward that goal (though perhaps illegally!). In general, the important point to note here is just that the representations underlying conative representational decision making are such that the organism can in fact make its behavior dependent on them.

Finally, I here need not—and therefore do not—commit myself to any particular account of the nature of neural computation or of exactly what it takes to explicitly store a given behavioral function: my account is consistent with different ways of spelling this out (see also Piccinini, 2015; Piccinini & Bahar, 2013; Piccinini & Scarantino, 2010). All that I need is that the account of computation chosen can make a distinction between computing a function and merely looking up the values of the function in a table. This, though, is true of all of the accounts of computation defended in the literature (Piccinini & Scarantino, 2011, 2010). (I return to this point in chapter 6.)

All in all: conative representational decision makers are organisms whose behavior depends on representations (or at least explicitly stored behavioral functions) about what they are to do. The content of these representations can be left open and need not involve cognitive representations, but it does need to be something that the organism can in fact act on.

Full Representational Decision Making

Full representational decision making, as I understand it here, is the combination of cognitive and conative representational decision making. Specifically, it concerns organisms that make decisions by consulting representations about what they are to do, and these representations, in turn, are phrased in terms of cognitive representations.

So, for example, an organism that cognitively represents the color of objects is a full representational decision maker if it acts on a rule like this one: "If the object is purple, blue, or green, do x; if the object is yellow, orange, or red, do y." Note that the organism thus acts on two kinds of representations: it acts on a conative representation—the conditional rule just cited—and cognitive representations—the representations of the colors of the relevant objects. However, nothing else here is different from the case of pure cognitive or pure conative representational decision making—it just combines these into one decision-making system.

One point that it is important to note when it comes to full representational decision making, though, is that it creates the potential for conflict among the two different decision-making systems that comprise it. In particular, since full representational decision making combines cognitive and conative representational decision making, it is possible that these two systems are still operative on their own and attempt to influence the organism's behavior singly and as part of the full representational decision-making system. Indeed, this turns out to be empirically quite plausible: for example, organisms can be torn between acting based on non-representationally triggered behavioral responses and acting based on conative representationally computed behavioral responses (Greene, 2008; Haidt, 2001; see also below). Given this, organisms might also need to find ways of adjudicating these conflicts among the different representational decision-making systems (and also among these and the non-representational decision-making systems). While there is much that could be said about how this can be accomplished, I will here sidestep this question and concentrate on the workings of the cognitive and conative representational decision-making systems in the absence of conflicts between them.

Before considering the empirical evidence for the reality of cognitive, conative, and full representational decision making, three final remarks concerning the nature of these decision-making systems are important. First, a terminological remark: there are a number of other terms that are often used to describe cognitive or conative representations—"beliefs," "desires," "preferences," "intentions," etc. I here avoid these terms, as different writers mean different things by them. For example, some writers distinguish "desires" from "preferences" (see, e.g., Pollock, 2006; Hausman, 2012; Schulz, 2015b), and some use them interchangeably (see, e.g., Sterelny, 2003; Sober & Wilson,

1998). To avoid confusion, therefore, I phrase my discussion in terms of cognitive and conative representations only.

Second, cognitive, conative, and full representational decision making is here understood to concern matters of cognitive architecture: it is about how the mind of the organism is organized. This is thus not a point about the kinds of explanations humans engage in as part of everyday life to make sense of each other's behaviors (see, e.g., Nichols & Stich, 2003; A. Goldman, 2006), and it is also not about useful ways of interpreting behavior in general (Dennett, 1987; Davidson, 1980).[13] Of course, it may be that people, even in ordinary life, refer to something like the cognitive and conative representations laid out here, and it may be that these kinds of representations are useful tools for understanding many different kinds of systems. However, these are separate issues from the ones that I am engaged with here, and so I will not consider them further.

Third, note that nothing in my characterization of representational decision making assumes that this has to be conscious. While some authors suggest that representational decision making has to be conscious (D. R. Griffin, 1984; Epstein, 1994), others deny this (Nichols & Stich, 2003; Carruthers, 2006). I here adopt the latter position, but that is largely because it is logically weaker: for all that follows, it is fine if it is argued that representational decision making has to be conscious—as long as it is admitted that many animals can be conscious in the required sense (see also D. R. Griffin, 1984).

The Reality of Representational Decision Making

It used to be controversial whether many (or indeed any) organisms should be seen to be representational decision makers (see, e.g., Stich, 1983; Brooks, 1991; van Gelder, 1995; Haugeland, 1999; see also Chemero, 2009; Ramsey, 2007; A. Morgan, 2014). In this section, I show why, by and large, this is no longer so—that is, why the consensus in the literature has moved toward acceptance of the reality of representational decision making.[14]

In particular, I consider three sources of evidence for the existence of representational decision making: human psychology and cognitive science, cognitive ethology and behavioral ecology, and neuroscience.[15] In each of these cases, the goal is not a complete overview of most of the relevant literature—a task that would take up several books all by itself—but just a summary of some key aspects of this literature. Also, the goal is not to note that some scientists sometimes make passing reference to representational decision making in their work; rather, the point is to show that seeing organismal decision making sometimes as representational makes for explanatorily and predictively successful theorizing.[16]

Human Psychology and Cognitive Science

In the human sciences of the mind and decision making, the widespread agreement on the reality of representational decision making comes out clearly by considering two areas: cognitive psychology and developmental psychology. Consider these in turn.

In cognitive psychology, the reality of representational decision making is a core commitment: most of the work in this area is about how mental representations—higher-level mental states intermediate between perception and action—function in a cognitive system. One example of this (though, as will be made clearer below, not an entirely uncontroversial one) is the large literature surrounding dual-systems theories of the mind (Epstein, 1994; Epstein et al., 1992; Chaiken & Trope, 1999; Kahneman, 2003; Stanovich, 2004; Sloman, 1996; T. D. Wilson, 2002; Clark, 2013; for a critical overview, see also Machery, 2009; J. S. Evans, 2008; J. S. Evans & Frankish, 2009).[17]

Dual-systems theory comes in many different variants (with often major differences between these variants). However, a core idea behind these different variants is that human decision making can be well understood as the result of two different psychological systems, (at least) one of which is a representational system.[18] (It is important to realize that, on some of accounts—see, e.g., Greene, 2008; Epstein, 1994—both systems are representational, so that the distinction between these two psychological systems should not be conflated with the distinction between representational and non-representational ways of making decisions; I return to this point below.) A relatively representative tabular characterization of these two systems is in Epstein (1994; see table 2.2).

While there is much more that can be said about the details of dual-systems models like this one, for present purposes, it is sufficient to note two key points.

On the one hand, there is no doubt that work based on dual-systems models of the mind is an empirically rich area of research. In particular, work

Table 2.2
Characterization of a Dual-Systems Cognitive Architecture

Experiential System	Rational System
Holistic	Analytic
Affective	Logical
Associationistic connections	Logical connections
Behavior mediated by "vibes" from past experiences	Behavior mediated by conscious appraisal of events
Encodes reality in concrete images, metaphors, and narratives	Encodes reality in abstract symbols, words, and numbers
More rapid processing	Slower processing
Slower to change	Changes more rapidly
More crudely differentiated: broad generalization gradient, stereotypical thinking	More highly differentiated
More crudely integrated	More highly integrated
Experienced passively and preconciously	Experienced actively and consciously
Self-evidently valid	Requires justification via logic and evidence

Source: Epstein (1994).

in this tradition has contributed significantly to our understanding of several major questions surrounding human decision making. So, for example, this literature predicts and confirms that forcing people to make decisions in less time leads to the quicker, associative system having a greater impact on their behavior, or that people will at times confabulate reasons for their behavior if that behavior is driven by the associative system (Haidt & Kesebir, 2010; Greene, 2008; Stanovich, 2004; Chaiken & Trope, 1999; Bago & De Neys, 2017).

On the other hand, while there is much debate about exactly how the two systems making up the models are to be characterized, what their inter-relations should be seen to be like, and how they generate behavior (Chaiken & Trope, 1999; Campbell & Kumar, 2012), it is clear that a commitment to a representational model of cognition—as it is understood here—is at the heart of this approach. Virtually all of the different versions of this model agree on the fact that (at least) one of the two systems of cognition is based on representations about states of the world and representations about behavioral goals or rules that tell the agent how to act on the world—that is, cognitive and conative representations (Greene, 2008; Chaiken & Trope, 1999; Epstein, 1994; Stanovich, 2004; T. D. Wilson, 2002; Ramsey, 2014; J. S. Evans & Frankish, 2009; Dolan & Dayan, 2013).

Now, as noted above, work on dual-systems theories of cognition is not uncontroversial. One important charge that has recently been leveled against this work is that the either / or distinction that is at its base is misleading (Kruglanski & Gigerenzer, 2011). According to Kruglanski and Gigerenzer (2011), a better way of understanding decision making is in terms of a number of "simple heuristics"—simple decision rules that are easy to apply and work with. Importantly, though, these simple heuristics can, in different contexts, have features of either of the two systems: for example, they can be unconscious or conscious, automatic or deliberate, and so on.[19]

There is no doubt that this attack on dual-systems models needs to be taken seriously, and there is also no doubt that more can be said about this appeal to simple heuristics (see chapter 8); however, for present purposes, the key point to note is that Kruglanski and Gigerenzer (2011) in fact do not attack the reality of representational decision making per se. In particular, they also see decision making as the result of the consultation of behavioral rules, and these rules are generally seen to compute over representations about the state of the world. So, while there are undoubtedly great differences between Kruglanski & Gigerenzer's (2011) theory and dual-systems models—especially concerning the details of how representational decision making should be seen to work—both sets of theories are in fact in agreement about the reality of representational decision making. For present purposes, this is all that matters: the goal here is not to defend dual-systems theory as such, but merely to note a commitment to the reality of representational decision making in cognitive psychology as a whole.

The second area of work in the human decision sciences (broadly understood) that it is useful to consider here is developmental psychology. The mainstream body of work in this area concerns itself with the investigation of the representations that human infants and children rely on when interacting with their environment (Hobbs & Spelke, 2015; de Hevia et al., 2014; Skerry et al., 2013; Gopnik & Schulz, 2004; Baillargeon et al., 2010; Carey, 2011; Nichols & Stich, 2003, chap. 2). Again, while there is much that could be said about the details of this work, for present purposes, it is sufficient to note that this is also a progressing and well-established body of work that is thoroughly committed to representational decision making: infants and children are assumed to react to how they represent the world to be and by consulting representations of what they ought to do to the world (Carey, 2011; Hobbs & Spelke, 2015; de Hevia et al., 2014; Skerry et al., 2013; Gopnik & Schulz, 2004). So, for example, this research provides evidence for the fact that preschoolers generate representations about which events cause which other events, and use these representations to intervene in the world (Gopnik & Schulz, 2004); similarly, there is evidence that attributing goals to others presupposes the ability to act on similar goals oneself (Skerry et al., 2013).

Now, it needs to be noted that there is also controversy surrounding this research. For example, Cecilia Heyes has argued that some of the research above assumes too quickly that infant cognition is based on innate representational systems, rather than on simpler mechanisms of associative learning (see, e.g., Heyes, 2014a, 2014b). While there is again much more that could be said about this dispute, for present purposes, the key point is to realize that it does not concern the reality of representational decision making either. Heyes is not denying that human children often make decisions using representations about what the world is like and what the world ought to be like (see, e.g., Heyes, 2013; Heyes & Frith, 2014). Rather, she is just arguing that (a) the content of these representations is often not culturally universal or innate, but learned with an associative learning process; and (b) children make representational decisions not nearly as often as is supposed in the mainstream research on this topic. For this reason, it is again best to see this dispute as concerning details about the way the representational decision-making system works, and not the outright reality of this way of making decisions.[20]

Therefore, despite the existence of some controversies, a deep and wide-ranging body of work in psychology and cognitive science suggests that much human decision making is representational. Consider next work on animal cognition.

Cognitive Ethology and Behavioral Ecology

The study of animal decision making has seen a "cognitive revolution" like the one that has occurred in the context of human decision making in the 1950s and 1960s (Tooby & Cosmides, 1992; Carruthers, 2002; Allen & Hauser, 1991; D. Jamieson & Bekoff, 1992; Allen, 2004). In particular, while there has been a long tradition of work that analyzed animal behavior without seeing animals as having rich mental lives of their own—perhaps because of worries about "anthropomorphizing" animals (Fisher, 1996)—many researchers in this area have now begun to take animal cognition more seriously (Allen, 2004; Dupre, 1996; Wilder, 1996; D. Jamieson & Bekoff, 1992). In particular, they have begun to apply the tools from cognitive science to the study of animal behavior, and have come to see animal decision making as resulting from psychological mechanisms that are very similar to the ones operating in humans. The reasons for this "cognitive turn" in ethology and behavioral ecology mirror those for the cognitive turn in the study of human decision making: the cognitive approach is both explanatorily and predictively superior to a purely behavior-focused approach. In particular, the cognitive approach can expand on and correct predictions and explanations derived from purely behaviorist models (D. Jamieson & Bekoff, 1992; W. J. Smith, 1990).

Importantly, furthermore, much of the work in this newer field of cognitive ethology is thoroughly representational in structure (see, e.g., Andrews, 2015; Allen, 2014; Allen & Bekoff, 1994; W. J. Smith, 1990; Allen & Bekoff, 1997; D. Jamieson & Bekoff, 1992). That is, much of this work analyzes some (though not all) animal decisions in terms of cognitive and conative representations as they have been laid out above—in fact, it is precisely the appeal to internal, representational mental states that marks this approach as so different from the behaviorist approach toward the study of animal behavior. There is a wealth of examples to underwrite this point—so much so that even a cursory overview is impossible—but the following list at least gives a sense of the role that cognitive and conative representations play in recent research in cognitive ethology:[21]

1. Russon (2003, pp. 298–302) argues that orangutan decision making is often based on representations about what the world is like (e.g., in terms of the properties of trees or foodstuffs); much the same goes for vervet monkeys and various other primates (Seyfarth & Cheney, 1992). Here, it is also worth noting that some of the researchers who deny that other primates think much like humans still think that these other primates are representational decision makers (see, e.g., Povinelli, 2003).

2. Ristau (1996) argues that some birds (like the piping plover) act on representations about what they want to achieve—namely, that a predator

follows them rather than remains at or near a nest site—and what they represent the world to be like—namely, what the predator actually does.[22]

3. Trimmer et al. (2008) argue that rats and other mammals are well seen to make decisions by relying on two systems akin to the dual-systems accounts proposed above: one system is effectively a "triggering" system (certain cues simply trigger behavioral outcomes), and one an information-processing system that leads to action only on the basis of representations about the world (such as the presence of predators).

4. Carruthers (2006, chap. 2) argues that some jumping spiders act on representations about the spatial layout of their environment (as well as about what other spiders can see), and combine these with representations about what they ought to do.

As just noted, many other examples exist, and there is much that could be said about the details of the above examples. However, for present purposes, this brief overview is enough: for while the details of the above examples may be controversial, and while it is important to be aware of the fact that much animal decision making may be non-representational (see, e.g., Kacelnik, 2012), there is no doubt that a significant body of research on animal cognition is both well-grounded and representational (in the way that is understood here).

Neuroscience

The study of the brain is now a central component of the study of cognition in general, and therefore overlaps a lot with the work on human and animal decision making mentioned above (E. Fehr & Camerer, 2007; Trimmer et al., 2008; Glimcher et al., 2005; Lieberman, 2010). However, there are several points about the research in (cognitive) neuroscience that are important to note separately here.

First, at least in mammalian brains, there are several brain regions that have been consistently linked to representational decision making. In particular, cognitive representational decision making is thought to especially involve a large cortical network comprising especially the left temporal lobe system (for object and face recognition), the inferior parietal lobe (for spatial cognition), and the dorsolateral prefrontal cortex (for planning, reasoning, and judgment in general; Goel & Dolan, 2003; Greene, 2008; Schroeder et al., 2010). So, for example, O'Reilly et al. (2014, p. 201) note that the neocortex (in general) "serves to integrate across experiences and extract statistical regularities that can be combinatorially combined to process novel situations." Similarly, conative representational decision making is associated especially with activation in the reward system of the brain (including, most prominently, the basal ganglia), as well as with activation in the medial and lateral prefrontal cortex and the anterior cingulate (for storage, recall, and evaluation of behavioral goals; Berkman & Lieberman, 2009; Greene, 2008; Schroeder, 2004; Morillo, 1990; Glimcher et al., 2005; J. W. Brown & Braver, 2005). So, for example, Berkman and Lieberman (2009, p. 103) note that a number of past studies "converge on the finding that dorsolateral PFC and parietal cortex are involved in representing the rules in maintaining extrinsic goals."

Second, it is important to note that some of the regions identified with representational decision making are—as noted in section III above—thought to also be involved with non-representational decision making (such as the basal ganglia). To understand this, two points need to be noted. On the one hand, this overlap should not be seen as surprising: after all, both the representational and the non-representational decision-making systems ultimately lead to behavioral outcomes, and thus have to be connected to the motor areas of the brain somehow. This implies that there will be some overlap between these two systems (see also Schroeder et al., 2010, p. 79; Moll et al., 2005). On the other hand, the neural overlap between the two decision-making systems is only partial. In particular, it is important to take into account which other systems are active: for example, while activation in the basal ganglia is associated with non-representational decision making if combined with activation in the amygdala or the hippocampus (Lieberman, 2003), it is associated with representational decision making if the ventromedial prefrontal cortex is activated as well (Schultz et al., 2000; Schroeder et al., 2010).

The third point to note here is that most of the work on the neural bases of representational decision making has concentrated on mammalian brains. Much less is known about the neural bases of representational decision making in avian, insect, and spider brains—though, as noted earlier, several researchers argue that many of these groups of organisms still contain representational decision makers. This is important to note, as it makes clear that we need to be careful in generalizing the findings from mammalian brains to other groups of animals. Unless we want to give up the assumption that different groups of animals could be representational decision makers in psychologically similar ways, we need to be open to the possibility that the neural realization of this trait in different groups of organisms is very different (for more on this, see, e.g., Shapiro, 2004). Still, the above research is useful for underwriting the reality of representational decision making in at least a large number of organisms (see also Colombo, 2014).

In short: cognitive neuroscience strongly underwrites the reality of representational forms of decision making. In particular, this work identifies several different neural regions associated with these different modes of interacting with the environment.

Therefore, there is a wealth of evidence supporting the reality of representational decision making in a wide variety of different organisms. While this does not mean that those denying the importance of representational decision making are completely misled—after all, the exact extent of representational decision making is still a matter of controversy—it does suggest that representational decision making is something that deserves to be taken seriously. However, this evidence does more than just underwrite the reality of representational decision making: it also makes clearer several further features of representational decision making.

General Features of Representational Decision Making

There are three features of representational decision making that much of the literature has come to agree on: the inferential nature of much representational decision making, the speed of representational decision making, and the high demand for cognitive resources of representational decision making. All three features are explored in more detail in chapters 4–6, but a preliminary statement of their wide acceptance is useful here to set the stage for that later discussion.

First, it is widely accepted that much representational decision making is based on representations that have been inferred from yet other representations, or uses mental representations as the basis for an inference to an appropriate form of behavior (Trimmer et al., 2008; Sterelny, 2003, chap. 5; Millikan, 2002, chap. 17; Papineau, 2003, chap. 3; Greene, 2008). While—given the way representational decision making is understood here—this is not as such *required* for a representational decision maker, it is still plausible that this is often the case. The reason for this is that the possibility of inference making allows organisms to be highly economical with their repertoire of stored mental representations or behavioral dispositions: they can just generate these representations or behavioral dispositions when they need to (see also Prinz, 2002, pp. 12–14).

Second, it is widely accepted that representational decision making is relatively slow when compared to non-representational decision making (Epstein, 1994; Greene, 2008; Trimmer et al., 2008; Fodor, 1983; Leventhal, 1982; Buck, 1991; Brewin, 1989; O'Reilly et al., 2014, p. 201). Part of the support for this claim comes from dual-systems accounts of cognition: according to these accounts, the more use a system makes of mental representations, the slower it is—with non-representational decision making at one end, and full representational decision making at the other (see, e.g., Leventhal, 1982; Buck, 1991; Brewin, 1989; Greene, 2008; Epstein, 1994; Ramsey, 2014). So, as noted in table 2.2 from Epstein (1994), the "associationistic" experiential system is said to lead to "more rapid processing," whereas the rational system is said to lead to "slower processing." A similar point is made by Fodor (1983, p. 64), who notes that "automatic responses are, in a certain sense, deeply unintelligent; of the whole range of computational (and, eventually, behavioral) options available to the organism, only a stereotyped subset is brought into play. But what you save by indulging in this sort of stupidity is not having to make up your mind, and making your mind up takes time. Reflexes, whatever their limitations, are not in jeopardy of being sicklied o'er with the pale cast of thought."

However, there are also some other theoretical reasons for thinking that representational decision making will generally be slower than non-representational decision making: since—as just noted—representational decision making often relies on representations or behaviors that are *inferred* from (other) representations, more cognitive labor needs to be expended as compared to non-representational decision making to yield a behavioral outcome. In turn, this is likely to make this kind of decision making slower than non-representational forms of decision making. Indeed, it is plausible that the more extensive the representational inferences are that an organism relies on, the slower the process of decision making will be: for example, an organism that relies on a lengthy representational inference (i.e., one involving many different steps, such as an inference from "it has been sunny every day for two weeks" to "the ground is very dry" to "digging a hole requires breaking through a tough layer of soil" to "it is necessary to obtain a tool to access foods stored underground") will be slower than one that relies on a shorter such inference (i.e., one involving fewer different steps, such as one from "it has been sunny every day for two weeks" to "the ground is very dry").

Third and finally, there is wide agreement on the fact that representational decision making often requires more in the way of cognitive resources like concentration and attention than non-representational decision making (Lieberman, 2003; Berkman & Lieberman, 2009; Greene, 2008).[23] That is, representational decision makers often need to concentrate more on their decision making, and their attention is often focused more strongly on the act of making decisions itself (as compared to non-representational decision makers).[24] Again, this is both empirically and theoretically plausible. On the empirical side, note, for example, that representational decision making tends to be more powerful—that is, tends to affect behavior more successfully—the more cognitive resources an organism has available. So Zaitchik et al. (2014, p. 161) write:

Recently, research has begun to focus on the acquisition of a set of domain general higher cognitive capacities known as executive function (EF) skills. This suite of EF skills includes working memory (WM), the capacity to inhibit competing responses, and the capacity to monitor and flexibly select among potentially relevant sources of information (set shifting). EF is implicated in the formation of abstract representations, which in turn aid flexibility in reasoning and rule following.

They also note (pp. 171–172):

A related framework . . . emphasizes the role of developing EF in the creation of active, abstract, symbolic, categorical, and relational representations (Morton & Munakata, 2002; Munakata, Snyder, & Chatham, 2012). In this view, development of the prefrontal cortex allows increasingly abstract representations to be maintained actively in WM. These active, abstract representations allow children to overcome responses based on latent representations of habitual or previously learned behaviors.

Theoretically, the inferential nature of representational decision making also should lead us to expect increased requirements for concentration and attention, since it is plausible that making inferences takes cognitive labor, and thus uses up cognitive resources. Indeed, it is plausible to think that the more complex the inferences are that a representational decision maker relies on, the more concentration and attention it takes to make the appropriate decisions (Greene, 2008; Epstein, 1994).

Conclusions

I have laid out a particular way of understanding representational decision making—that is, decision making based on mental content. Specifically, I have first argued that it is not necessary to commit to any particular
account of the nature of mental representation, as long as it has accepted
that there is such a thing as mental content. Second, I hope to have shown
that representational decision making is not a trivial trait that every organism should be credited with at all times: rather, many organisms should
be seen to be purely non-representational decision makers—that is, they
should be taken to interact with their environment by relying on mappings
between physical states (such as perceptual or neural states) considered just
as physical states, and particular forms of behavior.

On this basis, I have then laid out a picture of representational decision making that is empirically and theoretically well grounded and widely
accepted. According to this picture, full representational decision making
consists of the combination of two representational decision-making systems:
a cognitive and a conative system. The cognitive representational decision-
making system makes decisions by consulting multi-tracking representations
about states of the world (broadly understood). The conative representational
decision-making system makes decisions by consulting representations about
what the organism is to do (what its goals are). I have further shown that
research in psychology, cognitive ethology, and neuroscience provides good
support for the reality of these decision-making systems. Finally, I have
noted that representational decision making is widely accepted to often be
inferential, slower, and more demanding in terms of concentration and
attention than non-representational decision-making systems.

Having thus established that representational decision making is a widespread trait whose reality can be taken seriously, it now becomes possible
to consider its evolution. Before doing this, though, it is worthwhile to
consider the methodology of this kind of evolutionary psychology in more
detail. Doing this is the aim of the next chapter.

3   A Moderate Evolutionary Psychology

The goal of this book is the investigation of the evolution of representational decision making. However, there are a number of scholars who are very skeptical about the value and plausibility of projects like this one—that is, those that combine evolutionary biology, psychology, social science, and philosophy (see, e.g., Richardson, 2007; Buller, 2005). These scholars argue that such evolutionary psychological research (broadly understood) often rests only on poorly substantiated evolutionary biological speculations, and that it is therefore epistemically very unconvincing. My goal in this chapter is to respond to these skeptical concerns.[1]

More specifically, I here defend a moderate, evidential form of evolutionary psychology that is not prey to the concerns raised by the critics of other forms of evolutionary psychology, and which can still add much of value to discussions in psychology, social science, and philosophy. This moderate form is arrived at, on the one hand, through scaling back the aims of the project—namely, toward providing only partial evolutionary biological analyses of a given psychological phenomenon—and, on the other, through showing how a compelling but moderate form of evolutionary psychology can be gotten off the ground—namely, by using all of the relevant pieces of information about a given psychological trait.

The chapter is structured as follows. In the following section, I briefly lay out what I understand by evolutionary psychology. Next, I lay out an argument summarizing a number of key challenges to work in evolutionary psychology. I respond to this argument in the subsequent section. Then, I use this response to make clearer the kind of project this book is engaged in. Finally, I present my conclusions.

Evolutionary Psychology

Research in evolutionary psychology is extremely variegated in content and methods (see, e.g., Barkow et al., 1992; Buss & Hawley, 2010; Boyd & Richerson, 2005; Sterelny, 2003; Garson, 2014, chap. 3; Barrett, 2015).[2] Fortunately, for present purposes, a detailed examination of the differences between the different versions of evolutionary psychology is not necessary. All that is needed here is a brief statement of the nature of evolutionary psychology so that the rest of the discussion has a firm foundation. (This statement will also be further developed below.)

The common core of evolutionary psychological research is the appeal to evolutionary biology in order to make progress in our understanding of various psychological phenomena, such as the way human minds are structured or the way they work. Evolutionary psychologists typically start by identifying a given psychological trait—the propensity to detect cheaters in social exchanges (Cosmides & Tooby, 1992), say, or male human attitudes toward risk (Dekel & Scotchmer, 1999)—and then provide reasons for why this trait should be expected to evolve in the relevant set of environments (i.e., the environments that characterize the typical living conditions of the population of organisms in question). They furthermore tend to focus on natural selection as the driver of these evolutionary processes (though not necessarily so—Pinker, 1997, for example, argues that the human aesthetic sense has evolved as a by-product of other psychological capacities).

This sort of project is interesting for several different reasons. First and most obviously, it is inherently important: we care, for its own sake, about the evolutionary history of major psychological traits. Second, the evolutionary biological considerations can underwrite the idea that some or all humans now *have* the psychological trait in question (Machery, forthcoming). This is of major relevance in cases where the existence of the trait is controversial—as is true, for example, when it comes to the existence of psychological dispositions to detect cheaters in social exchanges (Buller, 2005; Cosmides & Tooby, 2008; Fodor, 2008).[3] Third, the evolutionary biological account can tell us more about how the psychological trait in question *operates*: for example, it might tell us something about the circumstances in which it is likely to have particularly pronounced effects on behavior, and the circumstances in which its effects on behavior are likely to be more muted. Finally, the evolutionary biological account can ground ascriptions about what the psychological trait is *for* (at least to the extent that the account is selective in nature; see Piccinini & Garson, 2014; Ariew et al., 2002; Allen et al., 1998). While there is much more that could be said about the details of work in evolutionary psychology, for now, this brief statement is all that is needed.

A Challenge to Evolutionary Psychology

Evolutionary psychology has been a popular field of research for several decades—both in the academic and in the non-academic world (see, e.g., Barkow et al., 1992; Buss & Hawley, 2010; Buss, 2014; Pinker, 1997; Barrett, 2015). However, since its inception, this field of research has also been extremely controversial, with a number of authors putting forward serious challenges to the plausibility of the work falling under this heading (see, e.g., Richardson, 2007; Buller, 2005; Lewontin, 1998; see also Garson, 2014, chap. 3). Given that the present book engages in a project broadly under the heading of "evolutionary psychology," it is thus important to address these challenges.

However, instead of trying to engage with all or most of these challenges—a task that is made hard by the fact that so many different challenges have been put forward—my aim here is more limited. Specifically, in what follows, my goal is just to present and discuss one argument that lays out a number of worries with work in evolutionary psychology. I do not pretend that this argument expresses all that people are concerned about when it comes to evolutionary psychological research;[4] however, I do think that the argument does a good job at summarizing some of the major issues that surround this kind of research. Moreover, I think that this argument is useful for making clearer what kind of project the rest of the book is engaged in.

This argument—which I will call the "anti-evolutionary psychology" (AEP) argument in what follows—can be stated like this:

The AEP Argument
1. Providing and substantiating a genuinely plausible, full evolutionary biological analysis of any trait is hard—and it is likely to be especially hard for psychological traits (Richardson, 2007; Buller, 2005; Kitcher, 1985; Shapiro, 2011b).
2. The practice of evolutionary psychology generally fails to do justice to the difficulties of providing genuinely plausible, full evolutionary biological

analyses—that is, it typically appears to be restricted to providing unsubstantiated evolutionary just-so stories for the psychological traits in question (Richardson, 2007; Garson, 2014, chap. 3).

3. So: the epistemic value of much research in evolutionary psychology is effectively nil.

Put more succinctly, the idea behind this argument is that research in evolutionary biology is hard, and that evolutionary psychologists often do not seem to even *try* to do justice to this fact (i.e., there seem to be few constraints on what makes for a plausible evolutionary psychological account). It is no wonder, then, that much research in evolutionary psychology is epistemically unconvincing, and should not be seen to count for much.

Before discussing this argument in more detail, it is important to note that, to my knowledge, no one has actually put this argument forward like that (at least in print). In this sense, therefore, the AEP argument is original to me. However, it is also important to note that I have assembled the argument from pieces that are readily available in the literature; in that sense, therefore, the AEP argument is not original to me and something that readers of the literature should be very familiar with. With that in mind, consider the premises and conclusions of this argument in more detail.

Conditions for a Complete and Compelling Evolutionary Biological Analysis

Premise (1) notes that, in order to provide a complete and compelling evolutionary biological account of any given trait, much work is necessary.[5] In particular, there are a number of conditions that would need to be satisfied before such an account could be considered plausible and empirically well substantiated. Unfortunately, the exact number and detailed content of these conditions is a matter of some controversy; fortunately, for present purposes, a detailed assessment of these conditions is not necessary. In fact, in the present context, it is sufficient to simply accept the following list of five conditions, drawn from Richardson (2007, pp. 99–104) and Brandon (1990, chap. 5):

(a) we need evidence about the ancestral state of the trait in question.

(b) we need evidence about the variability of the trait in question.

(c) we need evidence about the strength and source of selection for the trait in question.

(d) we need evidence about the structure of the population in question.

(e) we need evidence about the degree to which the trait in question is heritable.

Consider these five conditions in more detail.[6]

The first condition notes that we need to know what the ancestral state of the trait in question was in order to provide a compelling evolutionary biological analysis of this trait (Richardson, 2007, p. 104; Brandon, 1990, chap. 5; Sober, 2008, chap. 3). There are three main reasons for this. First, the ancestral state of a given trait has implications for the ease (i.e., the likelihood) with which this trait can evolve (Sober, 2008, chap. 3; Brandon, 1990, chap. 5). For a classic example, consider the evolution of powered flight. The probability with which the latter can evolve is likely to depend on whether the ancestral organism had the ability for non-powered flight (in birds, this is currently the most widely accepted answer; Xu et al., 2003) or whether powered flight evolved from an ancestor lacking the ability for non-powered flight (Dial et al., 2006).

Second, the ancestral state of the trait in question partially defines the speed with which this trait can spread through a population (Baum & Smith, 2013; Felsenstein, 2004; Sober, 2008). In particular, it may be that the evolution of powered flight from non-powered flight can proceed at a much faster rate than the evolution of powered flight from the absence of any flying abilities at all—say, because the latter requires more changes to the ancestral organism than the former. If so, then it is possible to estimate how long it should take for powered flight to spread through a population of organisms under the two conditions—which, in turn, can be used to date these evolutionary events as well as confirm the assignment of the ancestral state (Baum & Smith, 2013; Felsenstein, 2004).

Third, some authors draw a strong distinction between the first evolution of a trait—that is, the event where a trait first appears in a population and then spreads through it—and the maintenance of a given trait—when a trait is inherited from an ancestral organism and is not lost in the subsequent evolution of the population in question (Sober, 1984, 2008; West-Eberhard, 1992; Baum & Larson, 1991). (Sometimes, this distinction is used to mark a further distinction between traits that are "adaptations" and traits that are "exaptations"—see, e.g., Gould & Vrba, 1982.) However, this is the least important reason for considering the ancestral state of the organism, since a very strong distinction between the first evolution of a given trait and its later maintenance is in fact not plausible: in particular, generally, traits that cease to be under selection pressure cease to be maintained in the population. (Put differently, it is implausible to think that there is very strong "phylogenetic inertia" that keeps traits in a population unless they are forced out for some reason—for further discussion of this point, see, e.g., Baum & Smith, 2013; Felsenstein, 2004; Diaz-Uriarte & Garland, 1996; Harvey et al., 1995a; Harvey et al., 1995b; Reeve & Sherman, 1993; Wake, 1991.) For this reason, the distinction between the first evolution of a given trait and its later maintenance (as well as the distinction between "adaptation" and "exaptation") does not play a major role in the rest of this book.[7]

The second claim—condition (b) in the above list—concerns the variability of the trait in the population in question.[8] The reason for needing information about this is obvious: evolution requires variability, so unless there is variability, evolution will not take place (Godfrey-Smith, 2009; Sober, 2000). However, the point here is in fact more complex, since the extent of variability also determines *the way* in which biological evolution can proceed. Thus, if, in a population of organisms, having trait T' would be more adaptive than having trait T, then this need not imply that T' evolves, simply because the only alternative to T in the population may have been trait T", which is even less adaptive than T (see also Godfrey-Smith, 2001).

Condition (c) refers to the fact that natural selection—that is, the result of variation in heritable traits with fitness effects (Godfrey-Smith, 2009, chap. 2)—is a major, but not the only, determinant of the evolution of many traits (Godfrey-Smith, 2001; Orzack & Sober, 1994; Lewontin, 1970). Keeping this in mind matters, since traits under strong selection are likely to evolve to fixation (holding fixed the rest of [a]–[e] above), but traits under weak selection can get lost quite easily (Gillespie, 1998). In order to fully understand the evolution of a given trait, therefore, it is necessary to establish how strong the given selection pressures are. (This point is made even more important due to the fact that the selection pressures can change over time; in fact, they can even change because of the evolution of the very trait in question, as in frequency-dependent selection.)

Furthermore, and relatedly, we want to know why there is selection for a given trait (the "source laws," in the terminology of Sober, 2000). What makes it the case that the trait in question is under selection? Answering this question is important, in addition, because some traits with high fitness are in fact disadvantageous to the organism (these are traits—such as having genes producing some sickle-shaped hemoglobin in areas with a high incidence of malaria—that, while being disadvantageous by themselves, are correlated with traits that are advantageous; see also Sober, 2000, pp. 78–83). Note, furthermore, that, while answering this second question need not be necessary for answering the first question—it is possible to establish that a trait is under strong selection without being able to ascertain why that is (Gillespie, 1998)—answering this second question can still be very useful for answering the first question. At least sometimes, knowing why a trait is favored by natural selection can tell one, at least approximately, how much it is favored by the latter. This is especially so since organisms can change the selection pressures they are facing (Odling-Smee et al., 2003)—in which case understanding why a trait is selected for might in fact be necessary for ascertaining how long and how strongly it will be selected for.

Condition (d) states that a complete and convincing evolutionary biological analysis needs to address the population structure of the organisms in question. This is due to the fact the population structure can affect the direction and magnitude of the evolutionary changes that are occurring in the population (Brandon, 1990; Sober, 2008; Sober & Wilson, 1998). So, for example, if the population size is small, random factors—"drift"—play a large role in the evolution of any of the traits in that population (Gillespie, 1998). Also, if the population is subdivided into groups, various kinds of group-selectionist processes—in some sense of the term—can affect the way a given trait evolves (Sober & Wilson, 1998). Further, if there is a lot of migration into the population, then selection for a trait might be balanced or even swamped by the continuous arrival of organisms without the trait (Gillespie, 1998). Various other such options exist. Clearly, these points need to be addressed to get a full picture of the evolution of the trait in question.

Finally, condition (e) refers to the fact that it is necessary that a given trait is heritable for it to be able to evolve at all (Brandon, 1990; Godfrey-Smith, 2009). Traits that are only weakly heritable will not change drastically even if they are under strong selection. Moreover, traits might show a bias in their heritability (for example, organisms might tend to have offspring that are taller than they are); if so, then this bias might affect or even counteract other evolutionary processes (Brandon, 1990; Godfrey-Smith, 2009). Lastly, the basis of inheritance of a trait—that is, typically, the genes underlying the trait—can also have a profound impact on how this trait can evolve. For example, if the trait is genetically linked to other traits that are maladaptive, or if the trait is heterozygous and thus does not breed true, it might not evolve to fixation in a population even if it is under strong selection (Sober, 2000, pp. 125–130).

All of this makes clear that there is much that we need to know in order to provide a full evolutionary biological account of any given trait. However, the situation here is even more difficult for evolutionary psychology.

This is due to the fact that, while obtaining the needed knowledge to underwrite a full and compelling evolutionary biological analysis of a given trait is hard even in the best of cases (Richardson, 2007; Brandon, 1990), there are reasons to think that psychological traits will generally be far from the best of cases (Richardson, 2007; Buller, 2005; Sterelny, 2003, chap. 6). In particular, these kinds of traits do not fossilize; they are hard to assess in terms of their adaptive value (partially because they are hard to measure in the first place); and they are often species-specific (see, e.g., Sterelny & Griffiths, 1999). For these reasons, it seems very difficult to underwrite all of conditions (a)–(e) for these traits. In short: providing a full and compelling evolutionary biological analysis of psychological traits is likely to be especially difficult.

Evolutionary Just-So Story Telling in Evolutionary Psychology

The second premise of the AEP argument notes that a number of evolutionary psychological accounts seem to consist of nothing but unsubstantiated "just-so" storytelling (Richardson, 2007; Buller, 2005). More specifically, this premise notes that it is not just that much work in evolutionary psychology fails to provide a fully plausible, complete evolutionary biological account of the traits it investigates—rather, providing a fully plausible, complete evolutionary biological account often appears not to even be *attempted*. Instead, researchers in evolutionary psychology often seem to be content with providing some speculations as to why a given psychological trait might have evolved—and these speculations often appear not to be particularly well supported (Buller, 2005, chap. 3; Richardson, 2007, chap. 1).[9]

So, for example, consider the question of why humans are attracted to salty and fatty foods. A classic answer that is sometimes suggested (e.g., Pinker, 1997, pp. 207–208) is that these foods were rare in the evolutionary past of humans, and signaled high nutritional value: that is, those early humans that had a taste for salty and fatty foods had more energy—and thus a greater expected reproductive success—than those that did not. Hence, these tastes should now be taken for adaptations to detecting foods of high nutritional value—albeit adaptations that are no longer adaptive, as salty and fatty foods are now, in many cultures, highly abundant and of low nutritional value. Note that this account does not even consider whether and how food preferences in humans are heritable, what the ancestral state was of these preferences, what population structures humans evolved in, and what the alternative food preferences were. Moreover, little or no support is provided for the contention that having a taste for salty and fatty foods was in fact adaptive—perhaps salty and fatty foods never had much nutritional value, and these food preferences are merely accidental byproducts of genes (inherited from deep within the phylogenetic tree) that initiate the development of any sense of taste.[10]

Put differently, premise (2) of the AEP argument notes that there seem to be few constraints on acceptable evolutionary psychological accounts. In particular, it seems researchers are free to speculate about the reasons for why a given psychological trait spread in a given population—minimal standards of internal coherence and broad fit to some known facts seem to be all that is asked for.

The Lack of Epistemic Import of Evolutionary Psychological Research

Putting these two premises together leads to conclusion (3): there is not much epistemic value in most evolutionary psychological research. If providing a full and well-substantiated evolutionary biological account of some phenomenon is hard (by premise [1]), and if much evolutionary psychological research does not even attempt to live up to the standards of good evolutionary biological research (by premise [2]), then it seems that there is not much reason to take this kind of research seriously.

However, as I try to make clearer in what follows, I think this argument is ultimately unconvincing. In this, I am not alone (see, e.g., Machery & Barrett, 2006; Machery, forthcoming; Walter, 2009). However, my reasons for thinking that this argument is unconvincing are different from those of other critics of the argument. Bringing this out is the aim of the next section.

A Moderate Form of Evolutionary Psychology

The AEP argument goes wrong not so much in the substance of its assertions, but rather in the fact that it leaves out of contention options that should be considered. To show this, I shall reconsider this argument in the light of two further considerations: the value of partial evolutionary biological analyses, and the possibility to provide well-substantiated evolutionary psychological accounts. Consider these in turn in the following section.

As an aside, another problem that has been noted with the AEP argument is that it is not obvious whether it is really true that all of (a)–(e) (or whatever other set of conditions is put forward here) need to be satisfied for a compelling full evolutionary biological analysis of a given trait (see, e.g., Walter, 2009). In particular, it may be thought that requiring all of these conditions to be fully satisfied for an evolutionary biological account to be complete and convincing sets the bar too high: very little work would end up qualifying as providing complete and convincing evolutionary biological accounts. However, the situation here is unclear, in that Richardson (2007, pp. 107–108) has argued that this is not so, and that much work in evolutionary biology does satisfy these standards. Fortunately, assessing this dispute is not necessary here: my concerns with the AEP argument are independent of the plausibility of conditions (a)–(e).

Full and Partial Evolutionary Biological Analyses

The first problem with the AEP argument is that it presumes that a full evolutionary biological analysis must be the aim of all kinds of evolutionary biological research. However, this is not so. Instead, the aim of this kind of research can be considerably weaker: it might merely be to bring out *some* of the key elements affecting the evolution of a given trait. In other words: the goal might merely be the provision of one piece of the puzzle of the evolution of the trait in question—the assembly of the full puzzle might be left for another occasion.

To see why providing such a partial evolutionary biological analysis can still be inherently interesting and valuable, it is best to begin by returning to the above—trade book—example of the evolution of human food preferences for salty and fatty foods. For the sake of the argument, assume that Pinker has indeed provided good reason to think that there was selection for these food preferences in human evolutionary history, and that this selection was due to the fact that salty and fatty foods were of high nutritional value (I return to this assumption below). Moreover, continue to assume that we do not know the exact ancestral trait, the exact set of alternative traits, the heritability of human food preferences, and the exact structure and size of early human populations. Given all of this, should we conclude that knowing that there was selection for human food prefer-ences for salty and fatty foods—due to their leading to increased consump-tion of foods with high nutritional value—is useless and uninteresting? I think not. Knowing that there was selection for these food preferences due to their leading to an increased intake in food of high nutritional value is *one* of the things we want to know about the evolution of this trait. To be sure, it is only one of the things we want to know about the evolution of this trait—but this does not mean that, by itself, it has no epistemic value. More generally, there is epistemic interest in an account of the evolution of a given (psychological) trait that fails to satisfy all of the conditions on a full evolutionary biological analysis. This is so for two reasons.

First, while we may not be able to spell out in detail all of these condi-tions, we may be able to spell out enough of them in sufficient detail to give a sense of the overall picture of the evolution of the trait in question. So, we may be in a position to give a well-grounded account of the adaptive pres-sures impinging on a given trait in different sorts of environments, as well as to provide at least rough estimates for some of the remaining factors in (a)–(e) above. This matters, as the combination of rough estimates of some of the factors influencing the evolution of the trait in question plus knowledge—or at least a compelling account—of one of these factors can yield a reasonably good sense of the full evolutionary biological account here.

For example, assume that we knew food preferences are quite strongly heritable in other organisms; then that might give us reason to think they were also easily heritable in humans. Similarly, if we knew that food prefer-ences are quite variable in other animals (which is not unreasonable; see Lal-and & van Bergen, 2003), then that might give us reason to think that they were also quite variable among early humans. Furthermore, in some cases, the importance of natural selection vis-à-vis other evolutionary determinants can be estimated by assessing the kind of trait under investigation—in par-ticular, complex traits with large fitness benefits are fairly likely to evolve at least to a large extent by natural selection (Godfrey-Smith, 2001; Sterelny, 2003; Dawkins, 1986; see also Nilsson & Pelger, 1994). Altogether, this may at least provide a reasonable and well-grounded *suggestion* that the evolu-tion of food preferences in humans was driven quite strongly by natural selection in the way set out above—despite the fact that there is still much about this that we do not know and that we would want to confirm for a full account of this evolution.

Second, even if we cannot provide estimates for some of the remaining factors influencing the evolution of the trait in question, knowing about one (important) piece of the puzzle is still interesting. After all, this one piece still provides some information about the puzzle as a whole: even if we do not learn much else about the evolution of this trait, we still learn this one thing about it. This is especially relevant if the piece we learn something about is relatively central to the puzzle—which, as just noted, will often apply to knowledge about the selective pressures on a given trait. Note also that it is at least sometimes possible to know about one of these factors in isolation from knowing about the others: for example, an optimality analysis can provide evidence of the adaptiveness of a given trait in a certain kind of environment even in the absence of information about the heritability of the trait, its distribution across the phylogeny, or the structure of the population of organisms in question (A.I. Houston & McNamara, 1999; Orzack & Sober, 1994; Sober, 2008).

Now, it is important to note that in the background here is a kind of "epistemic separability" assumption: there is value in learning about one of the factors that can influence the evolution of a given trait in isolation from learning about others. This assumption is plausible, though, since the overall evolutionary trajectory of a trait is reasonably seen as a straightforward combination of the different factors that can influence this evolution (Godfrey-Smith, 2009).[11] So: while a trait under strong selection need not evolve to fixation in a given population if the population size is small (so that drift becomes very influential) or if there are counterbalancing biases in the heritability of this trait (among other things), it is still true that, in order to make sense of the fact that the trait is maintained at least in low frequencies in the population for a long period, we need to know about the selective pressures on the trait. In short, it is plausible to think that the assembly of a full evolutionary biological analysis of a given trait can be done in a piecemeal manner.

What all of this implies is the following. It may be true that providing a full evolutionary biological analysis of any given trait is hard; it may also be true that this is especially hard for psychological traits. However, it is not true that only a full evolutionary biological analysis of any given trait is worth having—by contrast, even a partial analysis can be useful and interesting. This is so, as such a partial analysis can be used with estimates of the other elements needed for a full analysis to at least obtain a sense of this full analysis, and at any rate, even by itself, such a partial analysis can still provide something worth knowing.[12]

Taking Evolutionary Psychology Seriously

The second problem with the AEP argument is that it presupposes that evolutionary psychological research must be based on unsubstantiated speculations only. However, this is not so: instead, it is at least sometimes possible to give good reasons for the claims that are being put forward under this heading. There are three aspects to this last point.

First, it is important not to jump from the fact that some work in evolutionary psychology is done poorly to the conclusion that all of it must be (see also Sober, 2000, pp. 27–28). It may be true that some evolutionary psychological "findings" are just speculations; however, this does not mean that all such findings need to be.

Second, many of the particular concerns surrounding the provision of evolutionary biological accounts of psychological or social scientific traits should not be overstated. While it is true that some psychological traits are species-specific, many are not (see, e.g., L. R. Santos & Rosati, 2015; and, as made clearer in chapter 2, representational decision making is one of the latter). Furthermore, it is possible to assess the adaptive value even of species-specific traits—for example, using optimality analyses (see, e.g., A.I. Houston & McNamara, 1999; Trimmer et al., 2011; Orzack & Sober, 1994; Maynard Smith, 1978). Finally, it is of course true that many psychological traits are hard to measure and—for this or other reasons—hard to assess in terms of adaptive value. However, this is not true for all psychological traits (and nor is it fundamentally different from physiological traits, such as coloration and gait). For example, the evolution of psychological traits can sometimes be assessed by considering the products of these traits, such as tools (which sometimes even fossilize—see, e.g., Mithen, 1990, for a good example of this). In short: it is not obvious that the evolutionary biological analysis of psychological traits must be drastically different from that of non-psychological traits.

Third (and relatedly), if done well, evolutionary psychological research will draw on all that we currently know and take to be relevant about a given psychological trait in order to arrive at appropriate conclusions about when and why it might be selected for (say). This is neither trivial nor necessarily impossible, and the result can be conclusions that are both surprising and well supported by a wide variety of different considerations (Maynard Smith, 1978). Of course, this is not to say that providing a plausible (partial) evolutionary account of a given psychological trait must always be easy. The point is just that there is no reason to think that it must always be impossibly hard. Put differently: there are at least some psychological traits that lend themselves to a (partial) evolutionary biological analysis. An example might make this clearer.[13]

Consider the evolution of behavioral innovation: the generation of new behavioral dispositions by an individual that can then spread through a population through social learning (Reader & Laland, 2003). Here is what we know about the ability for behavioral innovation: many different types of animals have this ability, but they do not display it equally frequently in all circumstances—animals typically innovate more when under stress or at the periphery of a social group (Reader & Laland, 2003). We also know something about the sorts of behaviors that are particular prone to be innovated—namely, foraging, tool use, and social interactions (Lee, 2003; Byrne, 2003; Laland & van Bergen, 2003). Finally, we know something about the costs of innovation—the potential for maladaptive behavior and various energetic costs (Laland & van Bergen, 2003; Lee, 2003).[14]

Given this, it is not just wild speculation to argue that the ability for behavioral innovation is adaptive for allowing organisms to increase the chances of fitness-enhancing interactions with the environment—but only in environments where the expected fitness of the existing behavioral repertoire falls below a certain threshold (Reader & Laland, 2003). For in the latter kinds of environments, the costs of behavioral innovation are likely to be quite low (as the risks of lowered fitness from behavioral experimentation are relatively low, given that the starting place is so low already), but the benefits quite large (Lee, 2003). Furthermore, given the fact that many different kinds of animals have this ability and the adaptive consequences of behavioral innovation are likely to be quite large, it is reasonable to think that these selective pressures will be an important part of any full evolutionary biological treatment of behavioral innovation.[15]

Now, it is of course true that the considerations this kind of analysis relies on may be mistaken in one way or another. However, this mere possibility of error should not be seen to invalidate the entire analysis: in virtually any area of investigation, there is always the possibility of a mistake or oversight. Short of Cartesian foundationalism, every epistemic project—in and outside of scientific contexts—has to countenance the possibility of mistakes; work in evolutionary psychology is not special in this regard. This can be made clearer by considering two other scientific methodologies: mathematical models and experimental studies.

Mathematical modeling is an extremely widely used method in the investigation of a large number of different issues in many different sciences, from ecology to economics (Weisberg, 2013; Humphreys, 2004; M. Morgan, 2012; Reiss & Frigg, 2009). Now, as has frequently been noted by critics of this kind of methodology, it is very easy to come up with a model that shows anything one wants to show: judicious choice of assumptions and modeling frameworks can virtually guarantee the derivation of any desired result (see, e.g., Sterelny & Griffiths, 1999, chap. 10.6; Weisberg, 2007; Sober, 2000, pp. 133–138; Cartwright, 1999; Schulz, 2015a). However, this does not mean that mathematical modeling is scientifically worthless; it just means that models need to be built on assumptions that, in one way or another, are informative about the phenomena to be modeled (Cartwright, 1999; Sober, 2000, pp. 133–138; Boyd & Richerson, 2005; Weisberg, 2007).[16] Of course, there is no guarantee that the assumptions chosen will indeed always be informative in this way—there is always the chance that a given scientific model is built on assumptions that are unconvincing for one reason or another. However, this does not mean that scientific modeling *in general* is problematic; it just means that scientific modeling is a fallible methodology that needs to be treated as such.

This point extends even to the seemingly most constrained, "hard-and-fast" part of science: experimental methods (for more on this, see, e.g., Gooding et al., 1989; Guala, 2005). Indeed, it might seem that, in an experiment, it is not possible to dictate what one will get, and that one is dependent on how the data end up turning out. However, this is misleading: in fact, experimental designs rely on good controls and compelling statistics—and it is not always obvious what these are. There is always the possibility that a particular experimental finding is due to some causal determinant that one has not controlled for, and the significance of a set of data often depends on how these data are statistically analyzed—for which there may be more than one option.[17] Moreover, experiments often yield a large number of different findings, and an assessment needs to be made as to which of these findings are important, and which can be ignored as being artifacts of the experimental setup. Again, therefore, we need to make a judgment about these issues.

However, just because this kind of judgment is involved, this does not make experimental methods unconvincing: it just means that these methods are fallible and that they need to be substantiated with appropriate foundations.

In short: work in evolutionary psychology is not fundamentally different from other parts of science. If done well, it will be based on a wide set of different considerations that it tries to balance in the most even-handed way possible. This does not guarantee that its conclusions will be true or unbiased—but this kind of guarantee is not available anywhere in science. In this way, it can be seen that the AEP argument goes wrong in assuming that all evolutionary psychological research needs to be unconstrained and "made up on the fly." In fact, this is not an intrinsic feature of this kind of research—like all scientific methodologies, it can be done well or badly, and the focus should be on good instances of this kind of work.

For an example of such a good instance, consider the research on psychological adaptations for cultural learning. There is now a set of well-developed models for when, why, and how it is adaptive to learn from others (Henrich, 2015; Boyd & Richerson, 2005); these models are also increasingly underwritten by research from anthropology and social psychology (Boyd et al., 2011; Henrich, 2015; Sterelny, 2012). Of course, questions remain about why psychological mechanisms facilitating cultural learning evolved—the models developed so far are only partial evolutionary accounts (e.g., we know little about the heritability of the psychological underpinnings for cultural learning or their exact past variability). Still, these accounts are now widely recognized as providing relatively strong evidence for the hypotheses that (a) some organisms (humans) have evolved specific psychological mechanisms for social learning, and (b) these mechanisms have certain specific features (e.g., they are cued toward successful individuals as sources of cultural learning: see Boyd & Richerson, 2005).

An Evidential Form of Evolutionary Psychology

Given these remarks, it becomes clear that the AEP argument is implausible as it stands. However, this does not mean that the strong forms of evolutionary psychology that many defenders of the approach tend to focus on have been vindicated. Instead, something more moderate emerges here: a purely evidential form of evolutionary psychology.

In particular, since I have granted that much evolutionary psychological research does not even attempt to provide a full evolutionary biological analysis of a given trait, the conclusion I reach is not that this research can fully confirm that a given trait is present in a given population (or that it is in the population for a given reason). This thus goes against remarks made by some evolutionary psychologists, who see the goal of their work as revolutionizing work in psychology and social science by creating a new science of the mind (Tooby & Cosmides, 1992; Buss, 2014). For the latter, we would indeed need the kind of complete and compelling evolutionary biological analyses sketched above—which, as made clearer by the AEP argument, we are unlikely to have.

However, this does not mean that the epistemic value of all evolutionary psychology is nil. Instead, a useful way of thinking about the implications of the above discussion is that evolutionary psychological research can provide *evidence* for given evolutionary biological, psychological, social scientific, or philosophical hypotheses. To understand this better, it is necessary to be clearer on the distinction between evidence and confirmation or acceptance (for more on this distinction, see Sober, 2008).

The confirmation or acceptance of a theoretical claim implies that enough evidence has been amassed to make this claim part of the accepted set of claims of contemporary science. There are many different ways of spelling out this idea—from Bayesian approaches that focus on the posterior probability of the claim to frequentist approaches that focus on the severity with which the claim has been tested—but they all have in common the idea that the outcome of the process is a (relatively) settled attitude toward the theoretical claim in question (Sober, 2008; Earman, 1992; Mayo, 1996; Howson & Urbach, 2006).

By contrast, evidential approaches just focus on the considerations at hand, and state that these considerations speak in favor of the theoretical claim in question (Royall, 1997; Goodman, 1999a, 1999b; Goodman & Royall, 1988; Sober, 2008). In other words, the idea behind evidential approaches is that we should restrict ourselves to noting that the relevant considerations *increase* our commitment to the theoretical claim in question—without going on to state that the increase needs to be toward full acceptance (or even that it needs to be very large). Indeed, we may have a good reason *not* to accept the claim: in the memorable example of Sober (2000), we may agree that a given noise is evidence for the fact that there are gremlins bowling in the attic, but decline to accept this last claim due the fact that we also have other evidence against it—such that gremlins are made-up creatures. There are again many different ways of spelling out this point: for example, we could see evidence as increasing the posterior probability in a hypothesis (given the data), though without determining from what to what (Schulz, 2011c), as involving a favorable likelihood comparison of the hypothesis relative to its competitors (Goodman & Royall, 1988; Royall, 1997; Sober, 2008), or as decreasing the imprecision of our credence in the hypothesis (Joyce, 2010). Fortunately, again, the details of this do not matter greatly for present purposes.

All that is important to note here is that evidential approaches have both strengths and weaknesses. The weakness is that evidential approaches depend on the fact that the hypotheses in question are well chosen: evidential approaches are (generally) comparative in nature, so that if the inputs to the comparison are flawed, the outputs will be as well. (The problem with taking the fact that we hear a given noise to be evidence for the hypothesis that there are gremlins bowling in the attic and against the hypothesis that it has started raining is that these two hypotheses make for a poor basis of comparison, given that we already have very strong evidence against one of them.) By contrast, the strength of evidential approaches is that they can add something of substance to the evaluation of (suitably chosen) hypotheses: without needing to consider all that can be said about a given set of hypotheses, they restrict themselves to stating whether a given set of considerations favors one (credible) hypothesis over another. (If we are unsure as to whether it is raining or snowing, then hearing a given noise may be very telling, given that snow normally does not make much sound when it falls. While there may be other reasons for why we hear the noise, in fact hearing it is one strike in favor of the claim that it is raining outside.)

Note that I do not want to argue that the evidence-based picture is the only right way of understanding all scientific inferences. All that I here want to point out is that evidence, though failing to lead to full confirmation or acceptance of a given theory, is valuable in and of itself and instrumentally. This is quite uncontroversial, though, in that other approaches to scientific inferences also accept that evidence matters (they might just want to go further and add other considerations to the latter as well; Sober, 2008).

Translating this point back into the present context, what this means is that much work in evolutionary psychology should be seen to take into account all that we know about a given psychological, social scientific, or philosophical phenomenon to present evolutionary biological considerations speaking in favor of a set of further hypotheses concerning this phenomenon. Put differently: the goal is to show that we have *some reason* to think that the trait in question will evolve in a relevant set of environments—for example (and typically), because it is adaptive for a specific reason. This can be made clearer by returning to the example of the human food preferences discussed earlier.

Continue to assume that we have good reason to think a taste for salty and fatty foods was adaptive in early human environments due to its leading to the ingestion of foods of high nutritional value. Given this, from an evidential point of view, the appropriate conclusion to draw is the following. The fact that these food preferences were adaptive for leading to increased eating of foods of high nutritional value gives us a reason to think that these food preferences evolved in humans, and that they did so because of their adaptive value. Furthermore, this claim can, in turn, be used as evidence for various other claims—such that humans will be led to eat foods of low nutritional value if the latter start to taste salty and fatty, or that a desire for salty and fatty foods will be particularly pronounced in humans that lack access to food of high nutritional value. Note that the idea is not that the adaptiveness of the food preferences *confirms* any of these claims—that is, that these claims should now be taken to be part of the accepted set of considerations of contemporary science. The idea is just that one point that speaks in favor of the evolution by natural selection of these human food preferences—and of the other hypotheses just mentioned—is that they were adaptive for leading early humans to eat foods of high nutritional value.

At this point, a skeptic of the value of evolutionary psychological research might argue that while it may be possible to obtain some kind of evidence about the factors that shaped the evolution of a given psychological trait (or related hypotheses), this does not mean that it is possible to obtain *strong* evidence about this. Put differently: a defender of the AEP argument might note that, given all of the unknowns that this research is based on, the evidence we can get about what drives the evolution of a given psychological trait is likely to be very weak only.

However, it is difficult to see why the evidence derived from work in evolutionary psychology *must* be weak: there is no reason for why it could not often be sufficiently strong to be taken seriously. At least, given that the goal here is only a partial evolutionary biological analysis, it is not clear why providing a well-grounded account of the selection pressures on a given psychological trait (say) must always be impossibly hard. Here it is also important to reiterate that evidence need not be overwhelmingly strong to be useful: there is no need to require that an evolutionary biological analysis of a psychological trait answers all questions that could be asked about this trait. Rather, the evolutionary biological analysis can just provide some evidence that, alongside other evidence, can help build up a fuller picture of the evolution and features of the trait in question.

A final point worth noting about this evidential way of understanding evolutionary psychology is that it should be distinguished from purely heuristic understandings of this research project. In particular, a number of authors (see, e.g., Machery, forthcoming; Schulz, 2012) have argued that the appeal to evolutionary biology in psychological research should frequently be understood to be useful just for suggesting hypotheses to investigate, experiments to do, and explanations to assess. However, by itself, the appeal should not be understood to add any epistemic weight to the hypotheses derived from it.

Now, it is important to realize that, while nothing in what I have said invalidates the heuristic understanding of research in evolutionary psychology, my defense of this research here is stronger than that. In particular, the claim here is that the evolutionary biological foundations of this research program can be more than just heuristically useful: they can be evidentially useful as well. While this does not amount to seeing evolutionary biology as revolutionizing the sciences of the mind, it is more than seeing evolutionary biology as providing a "thinking tool" for psychologists, social scientists, and philosophers. Put differently: there is no reason to restrict evolutionary psychology to a purely heuristic research program: it can be evidential in nature as well.

Summarizing the points of this section, therefore: the AEP argument goes wrong for two reasons. First, it presumes that the only kind of evolutionary biological analysis worth doing is a full one; however, this overlooks the fact that there is also much interest in a partial evolutionary biological analysis. Second, it presumes that research in evolutionary psychology is always poorly supported; however, this overlooks the fact that, if done well, research in evolutionary psychology will take into account all that we know and take to be relevant concerning a given phenomenon. In turn, what this implies is that work in evolutionary psychology can provide evidence for a given set of evolutionary biological, psychological, social scientific, or philosophical hypotheses (though it cannot fully confirm this set of hypotheses). While this last conclusion does not vindicate the extensive and central importance of work in evolutionary psychology advocated by some defenders of this kind of research, it is still a robust defense of this kind of work that deserves to be taken seriously as such.

Evidentially Investigating the Evolution of Representational Decision Making

The defense of the existence of a moderate, evidential form of evolutionary psychology is doubly relevant to the project of this book. In the first place, it provides some of the theoretical foundations for this project. Specifically, in what follows, I lay out an account of some of the key adaptive pressures that lead to the evolution of representational decision making. As will be made clearer throughout, the goal in this is not to pretend that I have settled all the questions that can be asked about this issue—that is, I do not pretend to have provided a full evolutionary biological analysis of representational decision making. All that I aim to do is to present well-founded considerations—that is, considerations that take into account the relevant known facts about representational decision making and the way evolution works in general—that speak in favor of the evolution of this trait, at least in some environments. That is, I try to present *evidence* for the claim that representational decision making (which, as noted in chapter 2, is relatively well understood both psychologically and neuroscientifically) has evolved in certain environments, and that it has done so for a specific set of reasons. As I hope the discussion in this chapter makes clearer, doing this is a worthwhile project, even if it does not attempt to settle all questions about the evolution of representational decision making.

Moreover, I think that there are some good reasons to think that the adaptive pressures on representational decision making are relatively important for the evolution of this trait. As noted in chapter 2, I think this trait is widely distributed across many different organisms and thus has had a significant period of time to evolve. Furthermore—as will be also made clearer in chapter 6—it is a complex trait requiring a number of changes to a nonrepresentational decision maker's cognitive architecture. For this reason, selective explanations of the evolution of this trait should be thought quite plausible (Godfrey-Smith, 2001; Dawkins, 1986; Nilsson & Pelger, 1994).

The other reason for why the present defense of a moderate, evidential form of evolutionary psychology is important is a converse of the first reason. In particular, the rest of the book can be read as an "existence proof" of the kind of evolutionary psychological research argued for in this chapter: it can be seen to make the arguments and conclusions of this chapter more concrete by showing that it is indeed possible to provide a well-grounded, partial evolutionary biological analysis of a psychological trait, and that doing so is inherently and instrumentally useful.

Conclusions

I have provided a defense of a moderate, evidential form of evolutionary psychology, according to which the goal of the analysis is just to provide evidence for a relevant set of evolutionary biological, psychological, social scientific, or philosophical hypotheses. I have tried to further make clear that this evidence derives from the fact that the analysis provides a partial evolutionary biological account of the relevant trait: using all of the tools and insights it has available to it, it lays out at least some of the major evolutionary pressures on that trait. While this falls short of providing a full account of the evolution of this trait, it is not epistemically worthless. In this way, I hope to have shown that there are forms of evolutionary psychological research that are immune from some of the major worries raised about this kind of research. In the rest of the book, I make this abstract defense of a moderate, evidential form of evolutionary psychology more concrete by applying it to the specific case of the evolution of representational decision making. To do this, I begin by considering some of the existing accounts of this evolution.

II   The Evolution of Representational Decision Making

4   The Need for a New Account of the Evolution of Representational Decision Making

A number of authors have inquired into the reasons why representational decision making might have evolved (though they have often considered this issue merely in the context of the investigation of other questions). The aim of this chapter is to analyze this work in more detail. More specifically, I try to show that while this work makes some very important suggestions that should not be overlooked, by itself, it fails to provide a detailed and well-grounded account of the evolution of representational decision making.

To bring this out, I discuss in some detail three of the key accounts of the evolution of representational decision making in the literature: (a) Ruth Millikan's (2002) specialization-based account, (b) causality-based accounts, and (c) Kim Sterelny's (2003) flexibility-based account. Concentrating on just these three accounts is plausible for two reasons.

On the one hand, since the remaining accounts (see, e.g., Godfrey-Smith, 1996; Kirsh, 1996; McFarland, 1996) often raise very similar issues to the ones discussed below, including these accounts here would not change the main conclusions.[1] On the other hand, the goal in what follows is not to show that there is no acceptable work on the evolution of representational decision making whatsoever. Rather, the goal is more constructive: while I am trying to substantiate that a new account of the evolution of representational decision making is needed, I also try to use the existing accounts to develop desiderata that a plausible account of the evolution of representational decision making needs to satisfy. Given this more constructive goal, a relatively more detailed discussion of fewer existing accounts is sufficient (and indeed preferable).

Before beginning this discussion, though, it is important to note that all of the accounts that this chapter concentrates on—just like the account defended in the rest of the book—see representational decision making as a distinctive trait with a selection-based evolutionary history of its own. As noted in chapter 3, this is not implausible given a moderate, evidential construal of the aims of these accounts. Put differently, the focus on selective explanations here should not be seen to rule out the possibilities that the evolution of representational decision making is entangled with the evolution of other traits (such as the ability to attribute mental states to other organisms, or to speak a language—see also Godfrey-Smith, 2002; Papineau, 2003, pp. 123–124), or that it has also been influenced by non-selective factors. It just means that a key aspect of the evolution of this trait is seen as being due to a distinctive set of selection pressures (see also Papineau, 2003, pp. 120–124). As made clearer in chapter 3, though, given the fact that representational decision making is a complex and (at least plausibly) adaptively highly important trait, this is quite plausible.

With this in mind, the rest of the chapter is structured as follows. In the first three sections below, I present and discuss accounts of the evolution of representational decision making from Ruth Millikan; defenders of the importance of causal reasoning; and Kim Sterelny. In the subsequent section, I bring out some of the upshots of this discussion that are useful to keep in mind in developing a novel account of the evolution of representational decision making. I summarize the discussion in the final section.

Millikan on the Evolution of Decision Making Based on Detached Descriptive and Directive Representations

According to Millikan, the evolution of representational decision making needs to be understood from the vantage point of its ancestral state: decision making based on what she calls "pushmi-pullyu" representations (P-P in what follows; Millikan, 2002). P-P are representational states that are simultaneously descriptive and directive: they both represent what the state of the world is like, and what an organism should do in this particular state of the world. Prototypical examples of P-P are beaver tail splashes (which simultaneously express that there is danger nearby and that other beavers ought to hide) and vervet monkey alarm calls (which simultaneously express that there is an eagle overhead, say, and that the monkeys nearby should hide underground). Genuine representational decision making—that is, decision making based on distinct cognitive and conative representations—then evolved by separating the two faces of P-P (Millikan, 2002).

Before considering this account of the evolution of representational decision making in more detail, it is useful to note that the trait Millikan considers to be the major alternative to representational decision making—decision making driven by P-P—is different from the one that is seen to be the major alternative to representational decision making in this book: namely, purely reflexive decision making. However, it turns out that the differences between these two contrast classes are not so important for present purposes.

Most importantly, this is because the reasons Millikan identifies for the evolution of cognitive and conative representational decision making are equally applicable (or at least applicable with only slight modifications) if the comparison is a purely non-representational, reflex-driven organism as if it is an organism driven by P-P. While Millikan argues that the evolution of specialized cognitive and conative representational decision making is an instance of a general evolutionary trend according to which new traits come about through the separation of different aspects of cruder ancestral traits (Millikan, 2002, p. 172), the rest of her arguments do not in fact depend on the truth of this specialization principle. Instead, they can just be read as providing reasons for the evolution of separate cognitive representational and conative representational decision-making systems. Put differently, her arguments lay out the benefits of cognitive representational and conative representational ways of making decisions, and these benefits are not entangled with the (presumed) P-P–based starting point of the relevant organisms.

(Of course, one might still wonder whether Millikan is right to focus on P-P–driven decision making as the relevant alternative to representational decision making. Here, it is worth recalling that, given Millikan's teleosemantic account of the nature of representation, P-P and reflexes—as they are understood here—need not be contrast classes: as noted in chapter 2, reflexes may be P-P in Millikan's sense. This is thus another reason not to take it to be important that Millikan focuses on P-P as the alternative to representational decision making, rather than purely reflexive decision making as is done here: it is not obvious how different these two really are.)

Why, then, does Millikan think separate cognitive representational and conative representational ways of making decisions are evolutionarily favored? In the main, she identifies three key pressures favoring the evolution of a distinct cognitive representational way of making decisions, and two key pressures favoring the evolution of a distinct conative representational way of making decisions.

The three key drivers of the evolution of cognitive representational decision making, according to Millikan, are (a) the benefits of a non-egocentric understanding of the objects in an organism's environments, (b) the benefits of a non-egocentric understanding of the spatial outlay of an organism's environment, and (c) the benefits of a non-egocentric understanding of the temporal conditional probabilities connecting various types of events in the organism's environment. In particular, Millikan argues that an organism can tailor its behavior more successfully to the world—both synchronically and diachronically—if it can recognize when it is interacting with the same object again, if it can go directly to relevant parts of its environment (e.g., where food is cached) without needing to rely on appropriate perceptions of the environment as a guide to trigger the appropriate bodily movements, and if it can anticipate what will happen after certain events occur (e.g., that there is erosion of the earth following heavy rain fall; Millikan, 2002, pp. 185–186, 187–188).

Millikan thinks that the benefits in (a)–(c) favor the reliance on cognitive representational states in decision making, since organisms just driven by P-P are unable to have the kinds of non-egocentric understandings of their objectual, spatial, or temporal environments noted in (a)–(c): P-P–driven organisms effectively have to react to every way of perceiving a given object, location, or event separately—they do not react to an object, location, or event per se, but only to the way the object, location, or event presents itself to the organism.[2] Because of this, they are unable to relate these different presentations to each other.

In turn, this matters, as it prevents organisms driven by P-P from streamlining their interactions with the environment, and from making these interactions more consistent with each other. For example, to realize that two different visual impressions are two different presentations of the same object, an organism needs the ability to at least represent shapes as such, or—even better—the whole object by itself (Millikan, 2002, pp. 172–173). Much the same goes for spatial and temporal cognition. Hence, over time, Millikan suggests that organisms will form and rely on object representations, spatial maps, and temporal maps in their interactions with their environment.

Switching to conative representational decision making, Millikan suggests that there are two features of this way of making decisions that make it adaptive relative to decision making driven by P-P. On the one hand, organisms driven purely by P-P can face the problem of engaging in overly stereotyped behaviors: for example, a goose may continue the motions of rolling an egg back into its nest even if the egg has fallen off the goose's bill (Millikan, 2002, p. 169). On the other hand, organisms driven purely by P-P are unable to find faster, more efficient, or otherwise better ways of reaching some goal state, even though they are generally capable of engaging in the relevant behaviors. For example, lovebirds carry the bark they need for building their nests under their wings, even though this is much less efficient than holding it in their beaks—which they are perfectly capable of doing (Millikan, 2002, p. 167).

These issues can be avoided, though, if the organism represents its goal states separately. For then it can monitor when it has achieved (or not achieved) a given goal state, and can thus alter its behavior accordingly. This avoids the problems of overly stereotyped behaviors (Millikan, 2002, p. 196). Also, organisms that represent their goal states separately can make inferences about how to best achieve these goal states: they can both reason forward from the state of the world currently and backward from their future goal state to find new ways of fulfilling these goal states. This avoids the problems of being stuck with sub-optimal behavioral reflexes (Millikan, 2002, pp. 204–207).

For these reasons, Millikan thinks that many organisms end up with specialized cognitive and conative representational decision-making systems. Note that, according to Millikan, these pressures toward the evolution of cognitive and conative representational decision making are separate from each other. In turn, this suggests that it is possible for the evolution of these two ways of making decisions to proceed at different rates: in particular, some organisms might—at least for a time—evolve only a cognitive representational decision-making system (see, e.g., Millikan, 2002, pp. 181, 191). That said, Millikan does seem to rule out the possibility that organisms evolve so as to be conative representational decision makers without first having evolved so as to be cognitive representational decision makers: representations of goal states, in order to play the above roles, need to be compared to representations of the current state of the world (see, e.g., Millikan, 2002, pp. 197–199). Hence, these other representations need to already be present in the mind of the organism. Still, overall, it is important to keep in mind that Millikan's account does not require that cognitive and conative representational decision making necessarily co-evolve.

However, when taking a step back and assessing the details of Millikan's account, it becomes clear that, while there is a wealth of insight in this account, it cannot be considered as laying out a fully plausible picture of the evolution of representational decision making.[3] There are three main reasons for this.

First, at key junctures, Millikan *hints* at what might drive the evolution of representational decision making without providing a detailed account of this. So, for example, Millikan does not make clear exactly why it would be so much more efficient for an organism to be able to recognize when it is interacting with the same object again, where it is located relative to a non-egocentric spatial map of its environment, or what is likely to happen after a given event occurs—and neither does she make clear exactly how great the benefits are that come from this. Similarly, Millikan does not make clear exactly why or how organisms can determine more easily how they can bring about their goals by reasoning to and from representations of goal states, or exactly how great the benefits are that come from this kind of reasoning. Of course, given the uncertainties involved here, a highly precise and empirically detailed account of these issues may not be in the cards. Still, it does seem that there is more to be said here than what Millikan provides (indeed, making this clearer is one of the aims of chapters 5 and 6 below). This point gains particular importance when combined with the second reason for doubting that Millikan has provided a fully plausible account of the evolution of representational decision making.

This second reason stems from the fact that Millikan does not consider that representational decision making also comes with costs. In particular, as noted in chapter 2, representational decision making is generally thought to be slower and to require more in the way of cognitive resources like concentration and attention. (Indeed, Millikan herself seems to acknowledge this point: see 2002, p. 208.) This matters, since it leads to an overstatement of the net benefits of representational decision making: if there are also costs that come from representational decision making, the actual (net) benefits of this way of interacting with the world may be lower than what Millikan thinks. Given that we know something about these costs, this should be taken into account, as it can affect the kinds and numbers of environments in which representational decision making should be expected to evolve.

The third problematic feature of Millikan's account concerns the fact that it does not pay sufficient attention to the abilities of organisms driven purely by P-P or reflexes. In particular, it is not so obvious that the latter could not also find ways of robustly tracking objects in their environment—in particular, they could rely on multiple cues arranged in a complex hierarchy or lattice (Sterelny, 2003, chap. 2; Millikan, 2002, pp. 166–167). The same holds for overly stereotyped behaviors: these, too, could be avoided by relying on more complex hierarchies of P-P or reflexes (e.g., perception of egg outside of nest → balance egg on bill and roll it back to the nest; while engaging in this behavior, check periodically whether egg is still on bill; if not, balance it on bill again, and continue rolling it toward the nest). Finally, while it might be true that organisms relying on detached cognitive and conative representational states in making decisions can adjust their behavior to the relevant state of the world very quickly, it is also true that (as noted in chapter 2) organisms can change or acquire new P-P or reflexes very quickly (Mackintosh, 1994; Dickinson, 1985; Dickinson & Balleine, 2000; Heyes, 2013).[4]

All in all, what this means is that it is not obvious Millikan has identified, in sufficient detail, the features of representational decision making that have driven the evolution of this trait. Consider, therefore, causality-based accounts of the evolution of representational decision making.

Causality-Based Accounts of the Evolution of Representational Decision Making

Several authors—including Dickinson and Balleine (2000); Papineau (2003); Gopnik et al. (2004); and Dayan (2012)—have suggested that the evolution of representational decision making should be seen to be driven by the benefits that a specific kind of causal reasoning makes available to an organism.[5] More specifically, according to these authors, only representational decision makers can engage in non-egocentric causal reasoning—which matters, since this kind of reasoning yields benefits that makes it highly adaptive when compared to purely reflexive decision making. There are two such benefits that are particularly worthwhile to focus on here.

First, genuine causal reasoners can engage in radically novel actions: they can put together a number of items of practical knowledge—that is, cognitive and conative mental representations—in ways that go beyond what is initially contained in these items of practical knowledge (see, e.g., Papineau, 2003; Dayan, 2012). For example, a causal reasoner who has learned that loud noises scare away many scavenging animals, and who has also learned that the presence of meat attracts scavengers, can put the two together to infer that eating next to loud things can be used to keep scavengers away from its food. Importantly, this organism can draw this conclusion even if it has never eaten near loud things before: the mere fact that the organism understands the causal structure of the world and that of its own actions is enough for it to be able to come up with this novel action.

Second, by understanding the causal structure of the relevant parts of their environment, organisms can fine-tune their behavior very quickly to the state of the world, without needing much in the way of a prior learning history. So, for example, if there is a food reward behind a gate, if sometimes lever A and sometimes lever B opens the gate, and if an organism can tell which is the case (e.g., by tracing the connections between the gate and the lever), the organism can immediately decide to pull the right lever, even without any kind of special trial-and-error learning (see Dayan, 2012). Similarly, if an organism has learned that opening gate A and gate B brings a different kind of food reward (e.g., protein vs. carbohydrates), it can decide to selectively open only one of these gates if it also learns that one of the food rewards is better suited to satisfy its current needs (Dickinson & Balleine, 2000). Note that this benefit differs from the previous benefit, in that the actions involved here are not novel. Instead, what is special here is that an organism can determine very easily which of its many *existing* behavioral dispositions it is useful to engage in the present context.

Now, in order to obtain these two kinds of benefits—that is, in order to make these two kinds of causal inferences—an organism needs to have an understanding of (a) which events cause which other events, and (b) which events should be caused—that is, what the organism's goals are. In turn, this requires the organism to be a representational decision maker (Dickinson & Balleine, 2000): causal knowledge and an understanding of the organism's goals requires the ability to form and act on cognitive and conative representational states. In this way, we arrive at an argument for the evolution of representational decision making: since causal reasoning is adaptive, and since causal reasoning requires the ability to make representational decisions, there is also reason to think representational decision making is adaptive.

Before assessing these causality-based accounts of the evolution of representational decision making in more detail, three further points ought to be noted concerning them. First, it is somewhat controversial exactly which organisms ought to be seen to be causal reasoners—and thus, which ought to be seen to be representational decision makers. For example, Dickinson and Balleine (2000) argue that rats certainly do qualify, whereas Papineau (2003) denies this. Fortunately, assessing this dispute is not necessary in the present circumstances—all that matters here are the reasons for why representational decision making should be expected to evolve.

Second, note that it seems to be empirically plausible that an organism could in fact learn the needed (causal) facts that underwrite causal reasoning. In particular, by appealing to the framework of Bayes nets, Gopnik et al. (2004) provide a detailed account of how, both in principle and in practice, causal learning is possible in realistic settings. Again, though, this is not so important for present purposes, so I will not discuss this further here.

Finally, note that, according to the causality-based accounts, the evolution of cognitive and conative representational ways of making decisions is necessarily correlated (see also Dickinson & Balleine, 2000, pp. 199–200). In particular, causal knowledge—that is, knowledge about which events cause which other events—is not thought to be adaptively useful unless organisms can use this causal knowledge to better bring about their goals; in turn, this seems to require that organisms can represent their goal states. On the other hand, the ability to represent goal states is also not thought to be adaptive by itself, as without representations about how to causally bring about these goal states, goal state representations are taken to be unable to function as drivers of an organism's behavior.

With all of this in background, what, then, can be said concerning the plausibility of causality-based accounts of the evolution of representational decision making? For present purposes, four major points need to be noted.

First, the fact that these accounts require a correlation in the evolution of cognitive and conative representational states is not plausible (see also Sterelny, 2003). On the one hand, as noted in chapter 2, there are neuroscientific reasons for thinking that the cognitive and conative representational decision-making systems can function independently of each other, and that they have, in fact, evolved somewhat separately from each other (Schroeder et al., 2010). On the other hand, contrary to what the causality-based accounts seem to hold, it is not obvious why an organism could not evolve an understanding of the causal structure of the world that it uses to merely trigger given behavioral responses—that is, without representing its goal states. Similarly, it is conceivable that organisms find it useful to represent their goal states even though they are unable to reason causally—for example, they might use these representations of goal states in non-causal reasoning. (Arguably, Millikan's 2002 account underwrites both of these theoretical possibilities.)

In fact, this last point can be taken to be a worry concerning causality-based accounts in its own right. In particular, there is no obvious reason why representational reasoning should be seen to be restricted to causal reasoning, or that only causal reasoning should be expected to be adaptively important in an organism's decision making. More specifically, reasoning concerning social roles and other correlations or patterns in the world—not all of which need to be causal—can be highly adaptive as well. So, knowing that organism X holds a position of a high status in the relevant society might be useful for another organism Y, as it allows Y to infer how X will act in various circumstances. However, this inference is not causal: it is not that X's social status is causally responsible for X's behavior—it is just a good predictor of it (e.g., due to the fact that there are various features of X's psychology that are a common cause of both X's social status and its behavior; Sterelny, 2003). In short: there is little reason to think that the only adaptive pressures influencing the evolution of representational decision making derive from the ability of representational decision makers to engage in causal reasoning—other sorts of reasoning abilities should be expected to be important as well.

The third worry concerning the causality-based accounts of the evolution of representational decision making is that it is again not clear that these accounts have identified features of representational decision making that provide benefits to organisms that they could not get by non-representational decision making instead. In particular, as noted earlier, reflex-driven organisms can also fine-tune their behavioral responses to the state of the world—and that quite quickly: they can adjust their behavior by acquiring new reflexes, and they can learn which behavioral responses are rewarded in which circumstances.[6] This makes it unclear exactly how much better off representational decision makers are (on the causality-based accounts). (In fact, even some defenders of causality-based accounts note that the adaptive pressures from causal reasoning are relatively small: see, e.g., Papineau, 2003, p. 101.)

The fourth worry with causality-based accounts builds on the third one: namely, the fact that causality-based accounts—like the account from Millikan—do not consider explicitly the costs of representational decision making. In turn, this suggests that the benefits of causal reasoning might be less than those identified by the causality-based accounts—in fact, the costs might even, at least at times, swamp the benefits of representational decision making identified here.

All in all, therefore, it is not obvious that the causality-based accounts of the evolution of representational decision making have the resources to fully explain this evolution. That said, it does seem plausible that, in at least some cases, the ability to reason causally is something that adds adaptive benefits to representational decision making: it does seem very plausible that organisms that can reason causally are at least sometimes in a position to adjust their behavioral responses faster and more radically to a changed environment than purely reflexively driven organisms. However, by itself, this ability does not make for a plausible candidate for being the driver of the evolution of representational decision making: it fails to treat the evolution of cognitive representational and conative representational decision making separately from each other, it fails to do justice to the importance of non-causal ways of reasoning, it fails to do justice to the powers of non-representational decision making, and it fails to take into account the costs of representational decision making.

Sterelny's Account of the Evolution of Representational Decision Making

Sterelny's (2003) account bears some similarities to Millikan's account (as well as to those of Sterelny, 1999, 2001; Godfrey-Smith, 1996; Kirsh, 1996; McFarland, 1996; Allen, 1999), but its details are quite unique. In order to lay out these details, it is best to begin by noting that, in its very structure, Sterelny's account—like that of Millikan—avoids one of the problems of the causality-based account: namely, it treats the evolution of cognitive representational decision making separately from that of conative representational decision making. Consider these two aspects of Sterelny's account—the one concerning the evolution of cognitive representational decision making and the one concerning the evolution of conative representational decision making—in turn.[7]

When it comes to the evolution of cognitive representational decision making, Sterelny thinks that the key feature to be explained is the fact that cognitive representations are "decoupled" from a particular behavioral response: they do not just trigger a given behavior, but inform many different behaviors. Slightly more specifically, according to Sterelny's account, the reason why organisms have evolved so as to rely on cognitive representational states—what he calls "decoupled representations"—in making decisions is that doing so can bring about a particular form of behavioral flexibility in the face of a certain kind of environmental complexity. More precisely, this part of his account rests on the following two theses.

Flexibility thesis: The ability to form decoupled representations allows organisms to react flexibly to the same state of the world.

Complexity thesis: Reacting flexibly to the same state of the world is adaptive in epistemically complex ("translucent") environments.

The flexibility thesis is meant to express the thought that—as also laid out in chapter 2—non-representationally driven organisms determine how to react to the state of the world by connecting a particular form of behavior to the presence of a given perceptual cue (Sterelny, 2003, pp. 27–29; see also Shapiro, 1999). However, according to Sterelny (2003), this implies that these non-representational decision makers are unable to be sensitive in their decision making to fine differences in how this perceptual cue came about and what other cues are present. By contrast, organisms that rely on decoupled representations in their decision making are not so constrained: since they decouple the registration of the occurrence of a given state of the world from any particular behavioral response to the occurrence of that state of the world, they can fine-tune their behavioral response to the precise conditions obtaining at the time of acting (Sterelny, 2003, chap. 3–4).

To this idea, Sterelny then adds the claim (expressed above in the complexity thesis) that the adaptiveness of acting on decoupled representations is a function of the epistemic complexity of the organism's environment. More specifically, the idea here is that in epistemically simple (or "transparent," in Sterelny's terminology) environments, the presence of various perceptual cues is sufficiently informative about the state of the world to allow the organism to connect, in an adaptive way, one behavioral response to each such feature. However, as these environments get more complex ("translucent," in Sterelny's terminology), these cues become less and less strongly correlated with the state of the world, thereby making it useful to vary the behavioral response to them (Sterelny, 2003, pp. 86, 92–93). Finally, in maximally epistemically complex ("opaque," in Sterelny's terminology) environments, there is no need to use decoupled representations either, as there, acting in line with the relevant objective probabilities is all that can be achieved: the organism cannot do better than to just rely on the probability distribution over the states of the world in determining its behavior. In short: according to Sterelny, the adaptiveness of acting on decoupled representations increases with the extent to which the organism's environment is epistemically complex, until it reaches a maximum, after which it drops off again.

Sterelny further argues that the major (though not necessarily the only) type of the kind of epistemically complex environment that favors decision making based on decoupled representations is a social one. In social environments, the intentions of other organisms are crucial for determining the best response to any given situation; however, determining these intentions, while not impossible, is not straightforward. For this reason, according to Sterelny, it is in particular social environments that favor the evolution of decision making based on decoupled representation, as it is particularly these environments that make it adaptive to fine-tune the organism's behavioral response to the exact state of the environment.

When it comes to the evolution of decision making based on conative representations—what Sterelny (2003) calls "preferences"—the situation is a bit different, however. In particular, Sterelny claims first that these kinds of representational mental states have the (evolutionary) function to carry information about the internal environment of the organism—that is, the state of its needs. However, since the internal environment of the organism is not epistemically complex (the organism's needs do not disguise or hide themselves, but signal their presence truthfully), he further argues that the evolution of preference-based decision making must have been driven by very different factors from those that drove the evolution of cognitive representational decision making.[8]

More specifically, Sterelny names the following four advantages of organisms that make decisions by relying on preferences (Sterelny, 2003, pp. 92–94).

1. Preference-driven organisms have an easier time making adaptive decisions when the range of behavioral options open to them is large. If the only behavioral outcomes open to an organism are "fight" or "flight," non-preference-driven decision making might be adequate; however, if there are n different types of fighting behaviors, and m different types of fleeing behaviors, it becomes much harder to see how a non-preference–driven organism can reliably act adaptively.

2. Preference-driven organisms do not need to rely on a vast number of motivational states. A sufficiently complex non-preference-driven organism would need an astronomically large number of drives (and other reflexes) in order to be able to behave adaptively. However, such a large number of drives is biologically implausible.[9]

3. Preference-driven organisms do not need to depend on unreliable mechanisms for adjudicating among their motivational states. Coordination among the many different active drives of a non-preference-driven organism can get difficult. In particular, according to one of the major mechanisms for achieving this coordination—the "winner take all" principle—the most "urgent" drive is given complete control over the organism's behavior; however, this kind of control mechanism will fail when it is necessary to take the urgings of other motivational states into account in order to behave adaptively (e.g., when drive to mate is adaptively combined with a drive not to get injured by suppressing mating cries during mating). Preference-driven organisms can avoid these sorts of problems.

4. Preference-driven organisms are able to cope more quickly with changes in their needs. Non-preference-driven organisms can only change their motivational structures in evolutionary time, whereas preference-driven organisms can learn what is good for them. This, though, will put the former at a disadvantage relative to the latter—they are less quick at adapting to their environment.

Putting the two sets of arguments—that is, the one concerning the evolution of decoupled representations and the one concerning the evolution of preferences—together, Sterelny argues that some environments (especially epistemically complex, relatively quickly changing ones) favor the evolution of representational decision making. However, both parts of Sterelny's account face some challenges.

The main problem with Sterelny's account of the evolution of decision making based on decoupled representations is that it fails to do justice to the powers of purely reflexive decision makers.[10] In particular, with suitably complex reflexes, a non-representationally driven organism can act just as adaptively as an organism that is driven by decoupled representations. This can be seen easily by noting that an organism can react flexibly—that is, non-uniformly—to some state of the world S by breaking it down into a number of other states of the world, each of which has a unique action associated with it. In this way, flexible responding to S can be realized with inflexible responding to sub-states of S (see also Grau, 2002, p. 77; Shettleworth, 2002, pp. 126–127; Pearce, 1994, pp. 123–128; Walsh, 1997; Sober, 1997).

Interestingly, Sterelny (2003, p. 35, footnote 3) seems to recognize this issue, but does not consider it to be problematic for his account. However, this is in fact the key point that needs to be addressed here: relying on decoupled representations is not the only (or most important) way in which an organism can respond flexibly to a complex environment—in fact, a purely reflexive responder might do so just as well (as least as far as Sterelny has shown). All that is needed is a finer individuation of the recognized states of the world.[11] Moreover, given the costs of cognitive representational decision making—which Sterelny does not take into account either—non-representational ways of making decision might in fact be more adaptive than cognitive representational ones.

When it comes to the second part of Sterelny's account—concerning the evolution of decision making based on preferences—there are two issues to be further considered. On the one hand, Sterelny does not make sufficiently clear why preference-based decision makers can avoid the problems he alleges to exist for non-preference-based decision makers. On the other hand, he does not make sufficiently clear why these non-preference-based decision makers have these problems in the first place. To see this, reconsider Sterelny's four alleged advantages of preference-based over non-preference-based decision making.

The main problem with Sterelny's alleged advantage 1 is that it is not spelled out in any kind of detail. It is just not clear what, exactly, the problem with "complex decision points" (Sterelny, 2003, pp. 92–93; see also 94–95) is meant to be for non-preference-based organisms, and so it is not clear how preferences could solve this problem. Why is it easier to figure out which actions one is to engage in if one can consult—using mental content—what one is to do than if one cannot? As will become clearer in chapter 6, I think Sterelny is gesturing at an important issue here; however, as it stands, this gesture cannot be considered a plausible argument. Much more work is needed to spell out this point.

When it comes to advantage 2, the problems are twofold. On the one hand, it is not at all clear why a large number of drives (and other non-representational motivational states) is biologically implausible—or, indeed, how many drives would be necessary to reach this number. In particular, it is hard to see how one is to determine or justify an absolute number of drives that is "too much"—indeed, it is not even clear what the right order of magnitude is here (5? 50? 500?). On the other hand and more importantly, it is not clear why the situation is meant to be better for preference-driven organisms: after all, they might also need a lot of preferences in order to be able to behave flexibly. Why is it more plausible to have many preferences than to have many drives?[12]

Advantage 3 suffers from the same sort of dual problem. On the one hand, it is not clear why a drive-based organism must rely on a "winner-take-all" mechanism to make decisions—after all, there is nothing intrinsic to drive-based organisms that makes this mechanism a necessity. For example, such an organism could also use a mechanism that gives different drives different weights in determining an action (a sort of "vector addition" model of action determination).[13] On the other hand, it is not clear why the same problem might not arise for preference-driven organisms as well; after all, these too have to find a way of coordinating their many different preferences. As a matter of fact, some of the most prominent accounts of how this coordination works are based on precisely the kind of "winner-take-all" principle that Sterelny thinks will often break down (see, e.g., Carruthers, 2006; Selfridge & Neisser, 1960).[14] Given this, it is not clear why preference-driven organisms are meant to be better off here than drive-based ones.[15]

Finally, the same applies to advantage 4. In the first place, it does not seem to be true that drive-based organisms need to be slow in changing their motivational states: as noted earlier, new drives are in fact often acquired very quickly and easily (Byrne, 2003; Mackintosh, 1983; Sherman et al., 1997, pp. 79–81; see also Dickinson & Balleine, 2000). On the other hand, it is not obvious that preference-driven organisms need to be much faster at acquiring new motivational states: in fact, it is at least conceivable that such organisms acquire most or all of their (fundamental) preferences either by (a) learning them in much the way that new drives are learned (see, e.g., Dickinson & Balleine, 2000), or (b) innately, so that they only change them over evolutionary time.[16] At any rate, there seems to be nothing inherent in preferences that make these possibilities more or less likely.

Overall, therefore, it becomes clear that Sterelny's account cannot be considered to provide a fully compelling picture of the evolution of representational decision making. In particular, it is not clear that suitably sophisticated non-representational organisms could not find ways of being just as flexible and successful in satisfying their needs as representationally driven organisms are. That said, Sterelny does point to some important insights: he hints at the need to consider how efficient representationally driven organisms are in relation to non-representationally driven ones, and he hints at the fact that these efficiency-differences may have something to do with the fact that representationally driven organisms can make inferences. However, these hints are exactly this—hints; they are not spelled out enough to be able to make the foundation for a fully convincing account of the evolution of representational decision making.

Desiderata for an Account of the Evolution of Representational Decision Making

The critical discussion of the existing accounts of the evolution of representational decision making is relevant here for two reasons. On the one hand, it shows that a new such account is needed: all of the existing accounts have problems that make them unconvincing as they stand. On the other hand, this discussion can be used to develop desiderata for a novel, plausible account of the evolution of representational decision making: it provides constraints that any such account has to satisfy in order to be plausible. In particular, there are four such desiderata that emerge out of the above discussion, described in turn below.

1. Respect the fact that cognitive and conative representational ways of making decisions can and plausibly have evolved separately.

The first desideratum notes that there are good theoretical and empirical reasons for considering the evolution of cognitive and conative representational ways of making decisions independently from each other—at least in the first instance. As was made clearer in the discussion of the causality-based accounts of the evolution of representational decision making, requiring that the evolution of these two ways of making decisions must necessarily—logically—be correlated is not plausible. Hence, any convincing account of the evolution of representational decision making must treat these two ways of making decisions separately from each other. Of course, this is entirely consistent with there being specific empirical or other reasons for thinking that the two are, in fact, often linked: for example, the evolution of one representational way of making decisions may make the evolution of the other easier, as some of the necessary representational machinery is already in place and just needs to be redeployed (I will return to this idea in chapter 6). However, the important point to note here is that it is reasonable to at least start by treating the evolution of cognitive representational ways of making decisions separately from that of conative representational ways of making decisions.

2. Consider the costs as well as benefits of representational decision making.

All three of the above accounts lack plausibility due to the fact that, while they may identify some benefits of representational decision making, they do not balance these benefits with the known costs of representational decision making. However, as noted in chapter 2 above, there are both theoretical and empirical reasons for thinking these costs are sufficiently substantial that they need to be taken seriously: representational decision making, due to its often inferential nature, is often likely to be slower and more cognitive resource-hungry than reflex-based decision making. A plausible account of the evolution of representational decision making needs to take this fact into account.

3. Provide a detailed account of the evolution of representational decision making.

The third desideratum is related to the second. A plausible account of the evolution of representational decision should specify—given the amount of detail that can be provided—exactly what the (net) benefits of representational decision making are, when these accrue, and to which organisms. That is, a plausible account of the evolution of representational decision making does not just say: "an adaptive advantage of representational decision making is X," but it will also say "the adaptive advantages of representational decision making are particularly pronounced in circumstances C1, C2, and C3, which thus makes it particularly plausible that organisms like O1, O2, and O3 have evolved this kind of decision making machinery."

4. Respect the powers of non-representationally driven organisms.

The last desideratum is one that also all of the above accounts struggled to satisfy. A plausible account of the evolution of representational decision making does not downplay the abilities of non-representational decision makers, but does justice to these. Put differently, a plausible account of the evolution of representational decision making does not lower the bar by assuming that non-representational decision makers are constrained in implausible ways; instead, it shows why representational decision making might evolve *despite* the fact that non-representational decision making is very powerful as well.

These four desiderata can be used as building blocks with which to construct a new account of the evolution of representational decision making.[17] Doing this is the aim of chapters 5 and 6.

Conclusions

I have considered some of the key existing accounts of the evolution of representational decision making: namely, Millikan's (2002) account, causality-based accounts, and Sterelny's (2003) account. Out of this consideration, two main conclusions have emerged. First, none of the existing accounts can be considered compelling as they stand: partly for idiosyncratic reasons, and partly for reasons common to these accounts, they all fall short of providing a convincing answer to the question of what drives the evolution of representational decision making. Second, though, they all also contain important insights that can be used as a springboard for the formulation of a new, more plausible account of the evolution of representational decision making. Indeed, I have used the discussion of these accounts to formulate four desiderata that an account of the evolution of representational decision making should satisfy in order to be considered plausible: it should consider the evolution of cognitive representational decision making and that of conative representational decision making separately (at least initially), it should consider the costs as well as the benefits of representational decision making, it should provide a detailed picture of the adaptive pressures shaping representational decision making, and it should do justice to the powers of non-representational decision makers. In the next two chapters, I lay out an account that tries to live up to these standards.

5   The Evolution of Cognitive Representational Decision Making

In this chapter, I develop a new account of the evolution of cognitive representational decision making—that is, of decision making that relies on representations about the state of the world. In line with the desiderata laid out in the previous chapter, this account will be independent of the evolution of conative representational decision making (though I will reconsider the relationship between the evolution of cognitive and conative representational decision making at the end of the chapter 6); it will be as detailed and precise as possible; it will pay attention to the costs of representational decision making; and it will do justice to the abilities of non-representational decision makers.

The core idea behind this account is that cognitive representational decision making can—at times—be more cognitively efficient than non-cognitive representational decision making. In particular, cognitive representational decision making, by being able to draw on the inferential resources of higher-level mental states, can enable organisms to adjust more easily to changes in their environment and to streamline their neural decision-making machinery (relative to non-representational decision makers). While these cognitive efficiency gains will sometimes be outweighed by the costs of this way of making decisions—that is, the fact that representational decision making is generally slower and more concentration- and attention-hungry than non-representational decision making—this will not always be the case. Moreover, it is possible to say in more detail which kinds of circumstances will favor the evolution of cognitive representational decision making, and which do not.[1]

The chapter is structured as follows. In the following section, I present the core idea of the account of the evolution of cognitive representational

decision making to be defended. Next, I add the costs of representational decision making to this account. Then, I use the insights of the previous two
sections to lay out the situations in which cognitive representational decision
making should, overall, be expected to evolve. Finally, I make clearer how
three issues of current concern in the literature on cognitive representational decision making (the gradedness of cognitive representations, the
social acquisition of cognitive representations, and the modularity of the
mind) can be incorporated into my account. I summarize the discussion in
the final section of this chapter.

Cognitive Representational Decision Making and its Adaptive Advantages

Cognitive representational decision making is, as noted in chapters 1 and 2,
a widespread, complex trait, and as such calls out for an explanation of its
evolution. In particular, we want to know which selective pressures favored
the evolution of this way of making decisions. To answer this question, it
is useful to break it into three parts. First, I lay out the *cognitive* benefits
of cognitive representational decision making. Given this, I then, second,
relate these cognitive benefits to the *adaptive* benefits of cognitive representational decision makers. Finally, I discuss some objections to this account
of the adaptive benefits of cognitive representational decision making.

Before getting started on laying out this account, though, it is important
to note an assumption in the background of the discussion to follow. In
this chapter, I contrast cognitive representational ways of making decisions
with purely non-representational ways of making decisions. However, this is
done for the sake of clarity only: all the conclusions arrived at in this chapter
carry over to the (empirically perhaps more plausible) situation in which the
ancestral organisms are conative representational decision makers—a point
that will be made clearer at the end of chapter 6.

Cognitive Benefits of Cognitive Representational Decision Making

As noted in chapter 2, non-representational decision-making systems are
based on a set of associations of perceptual states of an organism with
behaviors: the organism's "table of reflexes." In particular, the organism
uses the occurrence of perceptual states to non-representationally trigger
appropriate behavioral responses. So, for example, the olfactory detection

Table 5.1
A Robustly Tracking Non-representational Decision Maker

Perceptual Cue	Behavior
Visual pattern 1	Behavior 1 (e.g., freeze in place)
Visual pattern 2	Behavior 2 (e.g., fly away)
Visual pattern 3	Behavior 2 (e.g., fly away)
Tactile pattern 1	Behavior 3 (e.g., initiate mating behavior)
Tactile pattern 2	Behavior 2
Tactile pattern 3	Behavior 1
Auditory pattern 1	Behavior 2
Auditory pattern 2	Behavior 3
Auditory pattern 3	Behavior 1

of a certain chemical compound directly initiates cleaning behavior among the *P. barbatus* cleaner ants from chapter 1. More abstractly, a non-representational organism's decision-making mechanism will be based on something like table 5.1 to manage its interactions with its environment (this is a slightly expanded version of table 2.1 in chapter 2).

For present purposes, the key point to note about this way of making decisions is that it is very likely to contain a lot of redundancy. In particular, purely reflex-driven organisms like this are very likely to react in the same way to many different perceptual cues: for example, in table 5.1, the organism engages in behavior 2 (i.e., flying away) when registering visual pattern 2, visual pattern 3, tactile pattern 2, or auditory pattern 1. The main reason for the existence of this redundancy is that it is highly adaptive for most organisms to "robustly track" many aspects of their environment (see also Sterelny, 2003, chap. 2).

"Robust tracking" is the ability to use many different resources to track the occurrence of a given state of the world, and thus to make adaptive decisions in a wide variety of situations. For non-representational decision makers, this requires relying on an array (which may be learned or innately specified) of different perceptual triggers for the same behavioral response (Sterelny, 2003, chap. 2). So, for example, a prairie dog might trigger a flight response if it sees a hawk-like shape overhead, *or* if it hears the characteristic warning call of a conspecific, *or* if it detects smell of type A, *or* . . . (see also Sterelny, 2003, chap. 2).

As Sterelny (2003) has shown, there is good reason to think that it is adaptive for many (though not all) organisms to be robust trackers. This is because many organisms cannot rely on the presence of exactly one perceptual cue to trigger an adaptive response: landmarks are not always equally visible, predators do not always make the same characteristic sound when approaching, and not all edible things can be easily smelled. One way in which to solve this problem is by relying on a disjunctive *collection* of perceptual cues: while the same visual impressions may not always be available to guide the way home, there may be other things that an organism can rely on when they are not (such as certain specific sounds or the pull of the earth's magnetic field). Similarly, while predators might not always announce their presence with a distinctive call sign, they might signal their whereabouts in other ways, too (such as certain smells, say). Also, edible foods might not always give off a distinctive smell, but they might still have a particular look, softness, or texture. In short: it is plausible that robustly tracking the environment is adaptive for many organisms—for many organisms need to reliably react to the occurrence of situations that cannot be reliably associated with a single perceptual trigger (Sterelny, 2003, chaps. 2–3).[2]

For these reasons, it is plausible that many purely reflex-driven organisms have a lot of redundancy in their table of reflexes: after all, it is likely to be adaptive for many purely reflex-driven organisms to be robust trackers, and, for them, being a robust tracker just means being sensitive to many perceptual states in bringing about a given form of behavior. Note that organisms that robustly track the state of the world in this way do not do so by cognitively *representing* the fact that this state of the world occurs (Sterelny, 2003, chap. 3; Millikan, 2002). Rather, they just use an array of interrelated perceptual states to make decisions: they are reflex-driven organisms with a lot of redundancy in their table of reflexes. This is important to note, for it is precisely this redundancy that a cognitive representational decision maker can avoid.

To see this, note that the ability to represent the state of the world can be seen as a way to "bundle" perceptual cues: the organism no longer needs to be sensitive to many different perceptual cues in order to be able to adaptively respond to the state of the world—it can just track that state of the world directly, by cognitively representing that it is occurring (Borensztajn et al., 2014, p. 181). Put differently: representational decision makers can pick up on the *patterns* inherent in the many ways the world presents itself to

Table 5.2
A Non-inferential Cognitive Representational Decision Maker

Perceptual Cue	Cognitive Representation	Behavior
Visual pattern 1 Tactile pattern 3 Auditory pattern 3	World is in S1 (e.g., predator is present but distracted)	Behavior 1 (e.g., freeze in place)
Visual pattern 2 Visual pattern 3 Tactile pattern 2 Auditory pattern 1	World is in S2 (e.g., predator is present and focused on hunting)	Behavior 2 (e.g., fly away)
Tactile pattern 1 Auditory pattern 2	World is in S3 (e.g., potential mate and no predator is present)	Behavior 3 (e.g., initiate mating behavior)

them—they can react to the state of the world *as* that state of the world.[3] In turn, this can make their decision-making processes drastically more efficient.

To bring this out in more detail, it is necessary to supplement this idea with a more detailed picture of how cognitive representations are generated. This is necessary, as the reliance on cognitive representations may, at first, merely seem to add a layer to the decision-making process, without though changing this process substantively. In particular, if an organism uses perceptual cues to trigger the tokening of a given cognitive representation, and then uses this cognitive representation to trigger a behavioral response, the decision-making process would seem not to have been made more efficient in any way: the organism's table of reflexes would have merely been transformed from something like table 5.1 to something like table 5.2.

However, it seems clear that an organism relying on table 5.2 is no more cognitively efficient than one just relying on table 5.1. (In fact, given the extra elements included in table 5.2 and the general costs of relying on mental representations, it is plausible that such an organism would be *less* cognitively efficient than one relying just on table 5.1.)

To see why the reliance on cognitive representations can, after all, streamline an organism's decision making machinery is by realizing that, by grouping different perceptual states together—that is, by picking up on patterns in these perceptual states—it becomes possible to *compute* the right response from the environment by combining different such groupings.

Recall that this is one of the widely accepted features of representational decision makers noted in chapter 2: they are able to rely on inferential connections among mental representations. In particular, cognitive representational decision makers can be sensitive not just to the occurrence of certain "primitive" states of the world, but also to various combinations of these states of the world. This is important, as given any kind of redundancy in how the primitive states of the world are tracked, combinations of these primitive states will feature redundancy on a much higher order of magnitude (see also O'Reilly et al., 2014, pp. 201–203).

To make this clearer, continue to assume that an organism robustly tracks the occurrence of a given state of the world by using a (perhaps even nested) set of perceptual cues to trigger various cognitive representations. This time, though, add to this the fact that the organism needs to respond, not just to the occurrence of a given state of the world, but also to various combinations of these primitive states of the world. So, to use a slightly different example, assume that the organism's set of behavioral responses needs to be sensitive not just to whether it is raining or sunny, but to whether it has been sunny for two days, it has been raining for two days, or it is alternately raining and sunny. Table 5.3 summarizes this.

Now, assuming that there is some redundancy in how the organism tracks whether it is raining or sunny, an adaptive, purely reflex-driven organism night need a very long table of perceptual cues to solve the problem in table 5.3. For example, assuming that robustly tracking rain or sunshine can be done with just two perceptual cues (one visual and one auditory), then they might need to rely on something like table 5.4.

By contrast, a cognitive representational decision maker can just rely on something like a combination of table 5.1 and table 5.3, as in table 5.5.

Table 5.3
A Cross-Temporal Decision Problem

State of the World Day 1	State of the World Day 2	Adaptive Behavioral Response
Sun	Rain	Forage
Sun	Sun	Relocate (too dry)
Rain	Sun	Forage
Rain	Rain	Relocate (too wet)

Table 5.4
A Purely Reflex-Driven Organism Solving a Cross-Temporal Decision Problem

Perceptual Cue Day 1	Perceptual Cue Day 2	Behavior
	Visual pattern 1	Relocate (too dry)
Visual pattern 1 (tracks sun)	Auditory pattern 1	Relocate (too dry)
	Visual pattern 2	Forage
	Auditory pattern 2	Forage
	Visual pattern 1	Relocate (too dry)
Auditory pattern 1 (tracks sun)	Auditory pattern 1	Relocate (too dry)
	Visual pattern 2	Forage
	Auditory pattern 2	Forage
	Visual pattern 1	Forage
Visual pattern 2 (tracks rain)	Auditory pattern 1	Forage
	Visual pattern 2	Relocate (too wet)
	Auditory pattern 2	Relocate (too wet)
	Visual pattern 1	Forage
Auditory pattern 2 (tracks rain)	Auditory pattern 1	Forage
	Visual pattern 2	Relocate (too wet)
	Auditory pattern 2	Relocate (too wet)

Note that table 5.5 is much shorter than table 5.4. Table 5.4 consists of two ways of tracking rain, and two ways of tracking sunshine, which, over two days of tracking, yields $4 \times 4 = 16$ different behavioral dispositions to store and consider. By contrast, table 5.5 consists of two ways of tracking rain and two ways of tracking sunshine on day 1; the results of this are then combined with another two ways of tracking rain and two ways of tracking sunshine on day 2 to yield the final decision: this yields a total of only $4 + 4 = 8$ different behavioral dispositions to consider and store. In this case, therefore, relying on cognitive representations can reduce the number of behavioral dispositions the organism needs to store by half. For more complex cases, the difference will be significantly greater.[4]

It is easy to see why cognitive representational decision makers can get away with fewer behavioral dispositions in a case like this. Effectively, they discard needless information by breaking the overall decision problem into two independent sub-problems that are easier to solve, and then

Table 5.5

An Inferential Cognitive Representational Decision Maker Solving a Cross-Temporal Decision Problem

Perceptual Cue Day 1	Cognitive Representation Day 1	Perceptual Cue Day 2	Cognitive Representation Day 2	Behavior
Visual pattern 1		Visual pattern 1	Sun	Relocate
		Auditory pattern 1		
	Sun			
Auditory pattern 1		Visual pattern 2	Rain	Forage
		Auditory pattern 2		
Visual pattern 2		Visual pattern 2	Rain	Relocate
		Auditory pattern 2		
	Rain			
Auditory pattern 2		Visual pattern 1	Sun	Forage
		Auditory pattern 1		

determine the solution to the overall problem by combining the solutions of the sub-problems (see also O'Reilly et al., 2014, pp. 202–203). By contrast, purely reflexively driven organisms solve the entire decision problem "from the ground up": they keep track of options (such as how rain was tracked on day 1) that the cognitive representational decision maker disregards. Put differently, cognitive representational decision makers "organize" the information they receive before they decide how to react to that information: they "chunk" the information into disparate units—the weather on day 1, the weather on day 2—and then make a decision based on these chunked cognitive representational units, not the information they receive directly (see also Borensztajn et al., 2014; O'Reilly et al., 2014; Ericsson & Charness, 1994; Anderson et al., 1998; Cherniak, 1986, 2012; Halford et al., 1998).[5]

This point also becomes clear from considering neural network representations of tables 5.4 and 5.5. In particular, table 5.4 can be seen to be equivalent to a network with eight input nodes—one for each of the perceptual cues the organism is sensitive to across the two days—and two output nodes (one for each of the two behavioral responses it might engage in). The input nodes are connected to the output nodes in such a way that different activation pairs of the former nodes yield a specific behavioral response (by assumption, the input nodes always activate in pairs across the two days). An example of this involving the perceptual cues of a visual

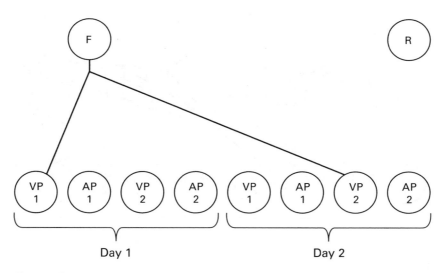

Figure 5.1
A network representation of an instance of the non-representational organism depicted in table 5.4 deciding to forage.

pattern of sunshine on day 1 and a visual pattern of rain on day 2 is in figure 5.1.

When putting all of these possible combinations together, this yields a network with 32 connections and 10 nodes, as in figure 5.2.

However, this contrasts radically with the network equivalent to table 5.5. This network interjects four cognitive representational intermediate nodes between the eight input nodes of figure 5.2, and then uses combinations of these intermediate nodes as triggers for the relevant behavioral outputs, as in figure 5.3.

Note that the intermediate nodes in this network—that is, the cognitive representations—function to separate the information the organism receives from the environment from the behavior it decides to engage in. In particular, the organism uses the information it has received from the environment to first form representations of what the weather is on the two days in question, and then uses combinations of the latter to decide what to do.

This is important, as it implies that the network in figure 5.3 is more streamlined than that of figure 5.2. In particular, while the network in figure 5.3 has slightly more nodes than that of figure 5.2 (14 vs. 10), it has radically fewer connections (16 vs. 32). Moreover, this latter difference increases quickly as

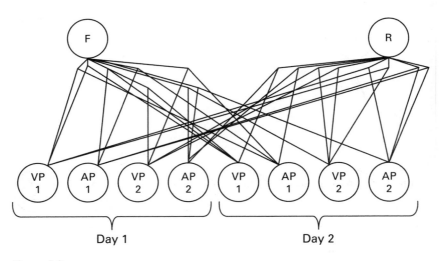

Figure 5.2
A network representation of table 5.4.

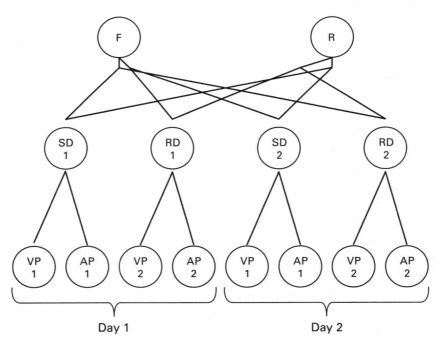

Figure 5.3
A network representation of table 5.5.

the number of perceptual cues an organism relies on to track a given state of the world increases (the former difference stays the same): for example, for an organism that uses three perceptual cues to track sunshine and rain, respectively, the number of connections for the non-representational decision maker increases to 72—six nodes with 12 connections each— whereas that of the cognitive representational decision maker increases just to 20: 12 connections to the intermediate nodes, plus 8 connections from the intermediate nodes to the output nodes. However, the difference in the absolute number of nodes in the two networks remains the same: four.

Importantly, these remarks about figures 5.2 and 5.3 and tables 5.4 and 5.5 should not be seen to be restricted to these figures only. Rather, these remarks illustrate and underwrite a point that is widely accepted in the literature: by exploiting structures in the behavioral dispositions of an organism, the reliance on cognitive representations can significantly streamline the way an organism makes decisions. So, as noted in chapter 2, O'Reilly et al. (2014, p. 201) also note that the neocortex (i.e., the neural regions underlying cognitive representational decision making) "serves to integrate across experiences and extract statistical regularities that can be combinatorially recombined to process novel situations." This is precisely the point made above. Related remarks can be found, for example, in Anderson et al. (1998); Cherniak (1986, 2012); and Ericsson and Charness (1994).

In a nutshell, therefore, the main point to take away from the discussion thus far is the following: by adding structure to an organism's perceptually driven behavioral dispositions, cognitive representational decision making allows for a more streamlined way of interacting with the environment as compared to non-representational decision making. The organism does not need to rely on as large a number of highly specific perceptual state-action connections, but can determine the appropriate behavioral response from a much smaller number of cognitive representation-action connections.

It is important to realize that implicit in this account is the fact that the relevant decision problem is in fact separable into a number of subproblems (or "chunks") that can be more easily solved by themselves (such as the weather on day 1 and the weather on day 2). There is no reason to think that this will be true for all decision problems: many decision problems might not break into such easily separable components as the weather on day 1 and the weather on day 2. However, importantly, it is also very plausible that many decision problems *are* separable in this way. In particular, this kind of separability is very plausible for decision problems that need to take into account spatial or temporal information—for this information is often separable in and of itself (e.g., a cross-spatial decision problem is likely to separable into different sub-problems that concern the different spatial regions concerned). Moreover, many decision problems are spatial or temporal in nature: in particular, problems such as deciding where food can be found or when it is safe to return to a given location after a predator has been spotted there are some of the most important problems of many different types of organisms. So, while there is no reason to think that every decision problem will profit from a cognitive representational approach, there is reason to think that many of them will (see also Prinz, 2002, chap. 11).

Relatedly, it is also important to recall (from chapter 2) that there are no a priori requirements or restrictions on the kinds of combinations of cognitive representations that an organism needs to be sensitive to for it to be a cognitive representational decision maker. In particular, there is no reason to assume that an organism must be able to form all and only logical combinations of its primitive cognitive representations. So, on the one hand, as in the above example, temporal, spatial, or causal representations are entirely possible as well. Further, it is unlikely that an organism can form all logical combinations of its primitive cognitive representations—there are very many of the latter, most of which will be irrelevant to the organism (see also Carruthers, 2006, pp. 78–81). Finally, there might even be reason to think that some organisms form cognitive representations in a way that directly contravenes basic logical principles—for example, they might infer from the cognitive representation that it is raining the further representation that it is raining and that there is a predator present nearby (see, e.g., Carey, 2011; de Hevia et al., 2014; Gopnik & Schulz, 2004).

However, that said, it is also important to note that the benefits of cognitive representational decision making come to the fore especially strongly the more such inferences can be made easily. A different way of putting this point is that while the present account does not require cognitive representational decision making to be highly systematic in nature, cognitive representational decision making's major strength comes out most clearly when systematic connections among mental representations can be appealed to. Again, this is quite in line with the literature on systematicity in general. So, Borensztajn et al. (2014, p. 181) note that mental representations at an intermediate level of complexity—what they call "encapsulated representations"—"provide the means for addressing systematicity since the relations between encapsulated representations are distinct from the relations between the lower-level inputs and are available to be used in information-processing operations." Similar points are made by Marcus (2014) and O'Reilly et al. (2014).

Taking stock, then: there are good reasons to think that cognitive representational decision making can streamline decision making by allowing the organism to react to grouped perceptual states, rather than to the perceptual states themselves. Put more crisply: cognitive representational decision makers can be more *cognitively efficient* than non-representational decision makers. While this is noteworthy in and of itself, for present purposes, the key question is why this kind of cognitive efficiency is biologically advantageous. What adaptive benefits are there in relying on a streamlined decision-making mechanism?

Adaptive Benefits of Cognitive Representational Decision Making

In one sense, the adaptive value of cognitive efficiency might appear not to need much further argument. G. C. Williams (1966), for example, considers the efficiency of a trait a hallmark of its being an adaptation—in fact, he notes that "economy and efficiency are universal characteristics of biological mechanisms" (G. C. Williams, 1966, p. 41). In this vein, the fact that the reliance on cognitive representations can streamline an organism's decision making machinery might, by itself, be seen to underwrite the biological advantageousness of cognitive representational decision making. Still, we would like some more specific reasons for why the cognitive efficiency of cognitive representational decision makers can be adaptively beneficial: *exactly why* does the reliance on cognitive representations increase the fitness of an organism? Fortunately, it is possible to answer this question; in particular, there are two reasons for why cognitive representational decision making can be expected to be adaptive.

The first of these reasons centers on the fact that the increased cognitive efficiency of cognitive representational decision makers enables these organisms to adjust more quickly and cheaply to changed environments. In particular, if the organism's environment changes, and the adaptive response to a given state of the world becomes different from what it used to be, a purely reflex-driven robust tracker needs to change a large number of behavioral dispositions—namely, all of those associated with perceptual states that track the relevant state of the world. By contrast, a cognitive representational decision maker only has to change one such disposition— namely the one concerning how it reacts to its *representing* the world to be in the relevant state. In turn, this is likely to make the adjustment to the new environment faster and cheaper.

To understand this better, note that it is overwhelmingly plausible that making any change to the behavioral dispositions of an organism (a) takes some time, and (b) involves some costs. Point (a) is true, since changing some of an organism's behavioral dispositions cannot be done instantaneously: the organism needs to acquire information about which behavioral dispositions need to be changed, it needs to acquire information about what these behavioral dispositions need to be changed to, and it needs to in fact make these changes. Given this, point (b) can also easily be seen to be true: on the one hand, the organism faces the opportunity costs of spending time making changes to its cognitive architecture—time that could be spent differently. On the other hand, the organism needs to expend energy on acquiring the needed information to make these changes and on actually making these changes.

Given (a) and (b), it is true that organisms are better off making fewer changes to their cognitive architecture rather than more (ceteris paribus; see also Simon, 1962; Carruthers, 2006, chap. 1). If it becomes adaptive to attack predators rather than freezing in place—say, because the target organism is rearing young—then an organism is better off if it can change just one cognitive representation-action connection, rather than having to change each of a large number of perceptual state-action connections surrounding the detection of a predator. Note that this is so even if the time and costs associated with any given change are quite low: for sufficiently large n (i.e., the number of perceptual cue-action connections), the product $r*n$ will be large even if r (i.e., the cost of changing one perceptual cue-action connection) is small. This matters, for it is indeed plausible that n is large for many organisms: after all, as just noted, it is plausible that there are often very many perceptual cues needed to track a given state of the world sufficiently robustly. Put differently: there is good reason to think that the increased cognitive efficiency of cognitive representational decision makers is adaptive also for enabling organisms to be more efficient at adjusting their behavior to changed environments.

The second reason for thinking the (sometime) cognitive efficiency of cognitive representational decision makers is adaptively important lies in the fact that there is reason to think the *cognitive* efficiency of a decision-making system is correlated with its *neurobiological* efficiency. There are several sources of evidence underwriting this point.

On the one hand, recent work on the neurobiology of concepts has suggested that theoretical concepts are arranged in a grid-like manner in the brain (Constantinescu et al., 2016; Hasselmo, 2012). So, for example, Constantinescu et al. (2016) trained human subjects to associate different Christmas symbols with different bird shapes, where the latter varied along two dimensions (neck length and leg length). When they then asked the subjects to recall these associations, fMRI scans revealed that these associations were underwritten by a neural grid matching the relevant table of conceptual associations: roughly, different bird shapes were represented by different neural regions, in a manner mirroring their conceptual relations. This suggests that more efficient structuring of the relevant *conceptual* space will make for a more efficient structuring of the related *neural* space as well. As the conceptual grid becomes smaller—for example, because it is no longer based purely on perceptual states (as in table 5.4), but now also includes cognitive representations that order these perceptual states in an efficient manner (as in table 5.5)—the neural real estate underlying this conceptual grid can become smaller as well.

On the other hand, there is now a considerable body of work suggesting that, as humans develop and increase their cognitive efficiency, their neurobiological architecture often gets simpler and more efficient. So, McGivern et al. (2002, p. 73) find support for the view that "increasing cognitive capacity during childhood may coincide with a gradual loss rather than formation of new synapses and presumably a strengthening of remaining synaptic connections." They also note that "during peak periods of synaptic proliferation in a region of cortex, information processing may be less efficient due to an excess number of synaptic contacts that have yet to be pruned. . . . Therefore, a decline in the functional efficiency of that region might be hypothesized due to a decrease in signal to noise ratio caused by the excess number of synapses" (McGivern et al., 2002, p. 74). Similar points are made by a large number of other authors (see, e.g., Casey et al., 2000; Neubauer et al., 2005; Neubauer & Fink, 2009; Grabner et al., 2004; Rypma et al., 2006; Sporns et al., 2000).

In fact, in the background here is a common and widely accepted finding of recent work in cognitive neuroscience: much cognitive maturation—which very plausibly includes increased cognitive efficiency—is associated with *neural pruning* (E. Santos & Noggle, 2011; Gazzaniga et al., 2009).[6] As mammals (at least) become more cognitively sophisticated, their brain structures often become more streamlined—synapses in the brain are culled to make the brain a more efficient processing system. So, E. Santos and Noggle (2011, p. 1464) note that by about age 2, "individuals are left with far more neurons and synapses than are functionally needed and/or preferred. Synaptic pruning is the process by which these extra synapses are eliminated thereby increasing the efficiency of the neural network." Of course, these findings only directly concern the development of neural efficiency *within* an organism. However, they also speak to the comparisons of the neural efficiency of different organisms, in that they show that cognitive efficiency, in general, often means relying on fewer synapses (not more). To make this clearer, it is important to forestall two possible misunderstandings about this point.

On the one hand, the point here is not that, as humans and other mammals get older, they have fewer neurons and synapses than when they were younger. The point is that, as humans and other mammals become more cognitively efficient, their neural system tends to become more streamlined. Of course, it is true that, as humans get older, they also learn much about the world—they acquire many novel cognitive (and conative) representations, for example—and hence it is plausible that their neural system does not get smaller (and might in fact get bigger) over time. However, what has been shown is that, at least across some periods of a person's life—for example, early infancy and adolescence—increases in cognitive efficiency by themselves (i.e., apart from the acquisition of new mental states of one kind or another) come with decreases in neural complexity. Put differently: increased efficiency in cognitive *processing* is correlated with increased neural efficiency.[7]

On the other hand, the point here is also not that that nonrepresentational decision makers are likely to run out of "neural real estate." Rather, the point is that it is metabolically costly to maintain a large brain, and that there are therefore selection pressures to keep the brain as streamlined as possible. This is similar to the venation structure in plant leaves: even though there is no reason to think that plants would run out of the materials that make up these veins, there still is reason to think that the assembly and maintenance of these materials is costly, and that there is therefore selection for efficient venation structure in plants (Roth-Nebelsick, 2001; Sack & Scoffoni, 2013). Or, to switch contexts: even if a country had enough funds available to expand its road network, it may not choose to do so, since any money spent on roads cannot be spent on other things—such as education or business development.

Putting all of this together: there are good reasons to think increasing the cognitive efficiency of an organism's decision-making system will also increase the neurobiological efficiency of that system. Note this does not mean that we need to expect the neural architecture of cognitive representational decision makers to directly mirror figure 5.3 or table 5.5. The point is just that, whatever the exact nature of the neural architecture underlying decision making, the reliance on cognitive representations can make this architecture more streamlined. Or, to put this point yet another way: recruiting the neocortical brain systems involved in cognitive representational decision making enables the organization of the brain to be more efficient overall than what would be true if the non-representational brain systems were expanded instead.

This matters, since it implies cognitive representational decision makers can gain some straightforward *biological* resources relative to non-representational decision makers: for example, their neurobiological efficiency increases the energy they have available for their remaining biological systems. So, the more neurally efficient cognitive representational decision makers can distribute energetic and other developmental resources away from decision making-related systems and toward other cognitive or non-cognitive systems: instead of maintaining a larger decision-making system, the organism can maintain a larger or more complex "executive control system"—the system that organizes the competing cognitive demands of the organism—or even just a larger body. Similarly, the organism can shift resources among different kinds of memory systems: an organism with a less extensive decision-making system can use some of the freed-up resources to increase its short-term memory (Klein et al., 2002; Tulving, 1985).[8]

All in all, therefore: there is good reason to think that the increased cognitive efficiency of cognitive representational decision makers is, at least in some circumstances, adaptive. On the one hand, this can make it easier for the organism to adjust to a changed environment, and on the other hand, this increased cognitive efficiency is likely to be associated with a more efficient neural organization as well. A few more points about this account are useful to note.

First, nothing in the above argument presumes that an organism that relies on cognitive representations *always* does so. In fact, the account is entirely consistent with only some decisions being based on cognitive representations, and the rest on reflexes. Everything here will depend on which decisions are most adaptively taken in which way—a point to which I return in the next two sections as well. In this way, the present account is perfectly in line with the point made in chapter 2 that no one presumes all organisms make all decisions using cognitive representations.

Second, it is important to note that the cognitive representational decision makers laid out in this chapter still rely on something like look-up tables. Put differently: in line with chapter 4, there are no conative representations that are assumed to underlie the organism's decision making apparatus. What makes the organisms at stake here representational decision makers is just the fact that the "triggers" of their actions are cognitive representational states. While, as will be made clearer in chapter 6, the fact that cognitive representations simply trigger a given behavioral response is not a necessity—cognitive representational decision makers can also rely on conative representations—what matters is just that purely cognitive representational decision making is at least a possibility. This is important to note, since it differs from the account of Sterelny (2003): as noted in chapter 4, at the core of his account is the idea that cognitive ("decoupled") representations enable the organism to react flexibly to the same state of the world. By contrast, on my account, this is not necessary: at the core of the account defended here is not the claim that cognitive representational organisms can do something non-representational organisms cannot (such as being a flexible responder), but that they can do the same things non-representational organisms can do—they just do them more efficiently.

Third, I do not want to claim that it must be the case that the only selective (or indeed evolutionary) pressures on cognitive representational decision making derive from the cognitive efficiency gains they make possible. Rather, it is consistent with my account that other such pressures exist, too: for example, as noted in chapter 4, it may well be plausible that cognitive representational decision makers are better at making causal inferences, and are therefore better at dealing with certain kinds of environments. The key point to note is just that there need not be a fundamental difference in the kinds of things cognitive representational decision makers can do (as compared to purely reflex-driven organisms) for them to be selected for: a major adaptive benefit of cognitive representational decision making lies in its allowing for more efficient behavior generation, and not in its allowing for the generation of new kinds of behaviors.

Objections

At this point, it is helpful to address two objections to the account laid out thus far in this chapter. These objections concern, on the one hand, the error profiles of cognitive representational and non-representational decision makers, and on the other hand, the breakdown patterns of these two ways of making decisions. Consider them in turn.

First, one might be concerned with the fact that the error profiles of the two decision-making systems appear to be different. When adjusting to a changed environment, cognitive representational decision makers have to make only one change, but if they get this wrong, many of their behavioral dispositions will be maladaptive. By contrast, purely reflexive decision makers have to make many changes, but even if they get one or a few of these changes wrong, the overall adaptiveness of their behavior is likely to still be quite high. If so, though, then it is not clear that cognitive representational decision makers are much better off.

This objection is not greatly problematic for the present account, though. This is so for two reasons. On the one hand, the force of this objection is not exactly clear, as it is hard to assess which of the two error profiles will be more adaptive when: much of this will depend on the specifics of the environment in question. On the other hand, whatever the outcome of this assessment, the overall upshot of the present account would remain unchanged anyway. This is because, in those cases in which the error profile of cognitive representational decision making is less adaptive than that of non-representational decision makers, then that is another cost that needs to be taken into account in the evolution of the latter (see the next section for more on this). By contrast, if the opposite is true, then this is another benefit of cognitive representational decision making that should be added to the benefits noted in this section. Either way, the overall conclusion of this chapter will not change drastically.

The second (and related) objection concerns the fact that cognitive representational decision makers might seem to show a more abrupt, less adaptive breakdown pattern. Breaking the connection between a cognitive representation and a given behavioral outcome completely disables the organism from acting in the relevant state of the world; by contrast, breaking the connection between a perceptual cue and a given behavioral outcome only leads the organism to react slightly sub-optimally in a given state of the world. For this reason, cognitive representational decision makers seem to be unable to display adaptive "graceful degradations" in function (this is a classic worry concerning symbolically representational decision making: see, e.g., Rumelhart et al., 1986).

However, this point is in fact misleading. Graceful degradation in function can also be had with cognitive representational decision makers. Indeed, the latter will show graceful degradation in function if the damage concerns the *tokening* of a given cognitive representation (in fact, given the latter's dependence on perceptual cues in generating cognitive representations, a very similar type of graceful degradation in function, as in the purely reflex-based case, can be expected here). Similarly, if the damage in the purely reflex-driven organism is a complete inability to recognize certain perceptual cues, these organisms will show a non-graceful degradation in function (whereas cognitive representational decision makers may, due to their inferential abilities, still be fairly functional).

All in all, what this implies is thus the following. Relying on cognitive representations when making decisions can streamline an organism's decision-making system in a specific way: by reacting to representations of states of the world—rather than just to perceptual states that track these states of the world—cognitive representational decision makers can get away with considering many fewer behavioral dispositions than reflex-based organisms have to. In turn, this can make cognitive representational decision making adaptive (see also Sober, 1998a; Whiten, 1995; Bennett, 1991; Tooby et al., 2005; Barrett, 2005). However, this is not where matters can be left: for there are also costs associated with cognitive representational decision making.

Adding the Costs of Cognitive Representational Decision Making

Pointing out the facts that cognitive representational decision makers can sometimes have an easier time adjusting to a changed environment and that they can have a more streamlined decision making machinery is only half the story, though. As made clear in chapters 2 and 4, (cognitive) representational decision making also comes with costs that need to be taken into account when reasoning about the evolution of this trait—in particular, relying on cognitive representations (especially when these are inferentially derived from other cognitive representations) will generally be slower and needier in terms of concentration and attention as compared to relying just on behavioral reflexes.[9]

Now, in a cost-benefit analysis like the present one, it is generally clear how these costs should be added to the account developed in the previous section: in particular, we ought to expect representational decision making to evolve only when the efficiency gains that come from this—the greater ease in adjusting to a changed environment and a more efficiently organized decision-making machinery—are greater than the efficiency losses that result from the slower and more concentration- and attention-hungry nature of representational decision making. However—and also in line with what was argued in chapter 4—we would like to go beyond these general statements to make more precise (at least as far as this is possible) when, exactly, we should expect the gains from cognitive representational decision making to outweigh its losses.

Fortunately, it is possible to do this. In particular, the above analysis suggests there are three main factors that determine the overall adaptiveness of cognitive representational decision making relative to purely reflex-based decision making.

First, one should expect cognitive representational decision making *not* to be adaptive when there is little need for an organism to robustly track a particular set of states of the world. This follows straight from what was said in the second section of this chapter: the benefits of cognitive representational decision making can only materialize if there is (significant) redundancy in the table of reflexes an organism relies on. Adding to this point the fact that cognitive representational decision making comes with costs suggests, in cases where organisms do not need to robustly track the state of the world, that the overall (net) balance of costs and benefits of cognitive representational decision making is likely to be negative.

This is important to note, as it is easy to describe some of the circumstances in which robust tracking is likely not to be adaptive. Obvious cases include the state of the organism's body: as noted above, for many organisms, relying on a single perceptual cue—the presence of pain—is enough to make attending to a particular body part adaptive; similarly, many organisms can rely on a single perceptual cue to decide whether to engage in feeding or drinking behavior—the presence of hunger or thirst (see also Sterelny, 2003, chap. 5). However, there are also some non-somatic states that might fall under this description: for example, some (reciprocal) helping behaviors can be adaptively driven by single perceptual cues. In particular, vampire bats seem to be able to adaptively make helping decisions just by attending to the begging behavior of colony fellows (see, e.g., Carter & Wilkinson, 2013; see also chapter 9 below). In general, therefore: cognitive representational decision making is unlikely to be adaptive when decision making based on single perceptual cues is adaptive (as is true for many bodily states).

The second case in which cognitive representational decision making is unlikely to be adaptive concerns situations in which speedy decision making that relies on few cognitive resources is of the essence. Cases like this are likely to make cognitive representational decision making maladaptive, since while cognitive representational decision makers still might feature significant efficiency gains here, the cognitive efficiency losses that come from their slower and concentration- and attention-hungry decision making machinery are likely to outweigh the former (however, see also chapter 8 below). Put differently: organisms are better off relying on a relatively inefficient decision-making system based on many perceptual cue-action connections that can be used quickly and without much need for concentration and attention if decision making time, concentration, or attention are in short supply.

Again, there are some obvious cases that fall into this category: for example, predator-prey interactions.[10] Many animals are better off relying on many perceptual cues to detect and react to the presence of a predator, rather than relying on cognitive representations—for a slight delay in decision making here can mean the difference between escape to safety and major injury or even death. In short: when very fast and frugal decision making is adaptive, cognitive representational decision making is likely not to be so. (See chapter 8 for more on this.)

The third situation in which cognitive representational decision making is unlikely to be adaptive concerns cases where there are few patterns to which the organism needs to react that can be constructed out of simpler cognitive representations. This point also follows straightforwardly from the second section above: even if organisms need to robustly track different states of the world, and even if fast and frugal decision making is not overly important in these cases, the efficiency gains from relying on cognitive representations are low when there is no way to construct these cognitive representations out of simpler ones. For in the latter case, the organism is in the situation of table 5.2 above: they may even become *less* efficient than purely reflex-driven organisms.

Which cases fall into this category? As noted earlier, an important set of these cases includes situations in which organisms do not need to—or cannot—take into account temporal, spatial, or causal *changes* in their behavioral responses. For example, if an organism just needs to make its behavior sensitive to whether it is raining—and not to how long it has been raining, or to what happened before it was raining—it can just robustly but non-representationally track rain, and is unlikely to benefit from cognitively representing that it is raining. Similarly, organisms that do not need to construct logical combinations of states of the world, but just the states themselves, would fall into this category. In particular, if an organism does not need to react to disjunctions or conjunctions of states of the world—say, that it is raining and that it is hot as opposed to that it is raining and that it is cold—they are unlikely to find cognitive representational decision making to be adaptive. In other words: cognitive representational decision making is unlikely to be adaptive for organisms that can react to fairly simple states of the world.

So, with all of this in mind, when *will* cognitive representational decision making be adaptive? By considering which situations are left out by the above three cases when cognitive representational decision making is likely *not* to be adaptive, it can be seen that cognitive representational decision making will be adaptive in situations that feature states of the world that need to be robustly tracked, combined with other states of the world to make adaptive decisions, and when decision making does not need to be especially fast and frugal. What are particular examples of these sorts of cases?

Environments in Which Cognitive Representational Decision Making Is Adaptive

In general, the key set of cases here concern the choice of which medium- and long-term strategies—that is, behaviors that extend across time and space—to pursue. Examples of these sorts of choices concern which mate to pick (especially for long-term partnerships—see, e.g., Buss & Schmitt, 1993), which foods to store so as to ensure nourishment when other sources of food are not available, which coalitions to join or build, how to navigate through a complex, extended terrain, and so on. These cases favor cognitive representational decision making, as extra decision making time, attention, or concentration are often not greatly important here—these decisions can often be taken relatively slowly and with much care—but where a vast variety of different factors in complex but (at least quite often) separable combinations often needs to be considered. For example, when plotting a route through a complex terrain to get closer to a large food source, organisms can often slowly consider several different options and compare them; nothing much is lost by this deliberation (they are not more likely to get injured or the food is not more likely to disappear). Furthermore, though, there is often still a large array of factors that need to be taken into account (are there any predators that need to be avoided? are there any food sources that could be tapped along the way? etc.).[11] Two further points are important to note about these kinds of cases.

First, organisms that live in a certain kind of complex social, spatial, or causal environment are likely to often encounter these situations. In these kinds of environments, adaptive behavior often depends on the appropriate choice of medium- to long-term strategies: in such environments, the fitness of an organism is heavily dependent on its joining the right coalitions, picking the right kind of mate, storing the right kinds of foods in the right kinds of places for the right kinds of time, and so on (Sterelny, 2003; Humphrey, 1986; Whiten & Byrne, 1997; A.I. Houston & McNamara, 1999; Colombo, 2014). In short: cognitive representational decision making is likely to often be adaptive in complex social, causal, or spatial environments.

The second point to note concerning the above cases when cognitive representational decision making is likely to be adaptive is that they (broadly) match the kinds of cases identified by the accounts of the evolution of representational decision making discussed in chapter 4. However, my account gets to this conclusion in a very different way from how these other authors get there. In particular, on my account, the reason these environments favor the evolution of cognitive representational decision making is not that cognitive representational decision making allows organisms to interact with their environment in a way that purely reflexive decision making does not. Rather (as noted earlier), the particular advantages of cognitive representational decision making—namely, their (overall) cognitive efficiency—come out particularly clearly in these environments. Put differently: my account and those of Millikan, Sterelny, Dickinson and Balleine, Papineau, and so on agree on the fact *that* cognitive representational decision making should be expected to evolve especially in complex social, spatial, or causal environments, but they differ over *why* cognitive representational decision making should be expected to evolve in those environments.

Overall, therefore, this all suggests that cognitive representational decision making is likely to evolve in cases (likely to be frequently encountered in complex social, spatial, or causal environments) when its gains—in terms of an easier adjustment to a changed environment and a more streamlined decision-making machinery—outweigh its losses in terms of decision-making speed and frugality. In this way, cognitive representational decision making can be more cognitively efficient *overall* than non-cognitive representational decision making: while it may be less efficient *at the moment of decision making* (due to its higher needs for time, concentration, and attention), it can be much more efficient *across time* (due to its less costly neural instantiation and adjustments to changed environments). Put differently: the dynamic efficiency gains of cognitive representational decision making can outweigh its static efficiency losses to make it more efficient overall than non-cognitive representational decision making.

Extensions

The remaining task of this chapter is to show how my account relates to various ongoing debates surrounding cognitive representational decision making. In particular, in what follows, I make clear how my account can be extended to cover (or how it can otherwise accommodate) three key issues discussed in the literature on representational decision making: the possible "gradedness" of many or all cognitive representations, the fact that many cognitive representations might be maladaptive due to their social origins, and the idea that (representational) cognition should be seen to be (massively) modular. Consider these issues in turn.

Graded Cognitive Representations

Much work surrounding cognitive representational decision making assumes that these mental states can be—or perhaps always are—*graded*. That is, it is often presumed that organisms often—or even always—should be seen to not just represent that the environment is in state S, but also to represent that the environment is in state S with differing degrees of strength (Savage, 1954; Jeffrey, 1983, 1992; Howson & Urbach, 2006; Joyce, 2009). For example, in many contexts, it is assumed that an organism should not (just) be seen to represent that it is raining, but also (at least sometimes) to represent that it is raining to degree 0.3.

A key reason for appealing to graded cognitive representations is that doing so is theoretically and empirically useful. In particular, economists, psychologists, and cognitive ethologists have found that decision making in both humans and other animals can often be well modeled by seeing an agent's cognitive representations as being graded—for example, on a probabilistic scale (see, e.g., Savage, 1954; Jeffrey, 1983; de Finetti, 1980; A.I. Houston & McNamara, 1999; Colombo & Hartmann, 2017) or on some non-probabilistic scale (see, e.g., Shafer, 1976; Dempster, 1968; Yager, 1987; Spohn, 2012). More specifically, it appears that, in cases where organisms receive evidence that does not unequivocally suggest that the state of the world is S, many organisms act as if they "shade down" their cognitive representation that the world is in state S. For example, they might be less inclined to do something that gets them a given food reward when the state of the world is S if their evidence for the fact that the world is in fact in state S is less than fully certainty-inducing (see, e.g., D. W. Stephens, 1989; Savage, 1954). A natural way of interpreting this is as seeing these organisms as only weakly committed to the representation that the world is in state S.[12]

This appeal to graded cognitive representations is important to consider here, as the account defended in this chapter is, as such, based on non-graded (purely qualitative) cognitive representations. For this reason, it might appear that it is not clear how my account can handle the theoretical and empirical work based on graded cognitive representations. Fortunately, it is in fact the case that my account *can* make sense of this work. Indeed, there are two different (but mutually compatible) ways in which it can do so, both of which draw on the fact that it is possible to mimic graded cognitive representations using non-graded cognitive representations (this is sometimes called a "Lockean thesis"—see, e.g., Foley, 1992; Hawthorne, 2009; Huber, 2009).[13]

First, organismic reactions to uncertain evidence can be modeled by seeing perceptual cues as triggering different cognitive representations only probabilistically. On this proposal, it is not the case that an organism always forms a given cognitive representation when presented with a given set of perceptual cues. Rather, there is a certain probability associated with the organism forming this cognitive representation in the face of this set of cues. In this way, while the organism still forms purely qualitative cognitive representations (for example, the organism might either represent the world as raining or not), the formation of these cognitive representations can be inherently graded: there is no guarantee that the organism will or will not represent the world as raining—all that we have is a probabilistic determination of the formation of this cognitive representation. Put differently: the claim that an organism represents that the world is in state S to degree 0.7 is here understood as the claim that, in the relevant type of situation, there is probability of 0.7 that the organism will form the representation that the world is in state S. Whether having such probabilistically triggered cognitive representations is adaptive depends on the details of the case (see, e.g., Cooper, 2001), but what matters here is just that, in principle, my account is perfectly consistent with this being the case.[14]

Second, it is possible that the content of a cognitive representation refers to matters of degree. In particular, it is possible that organisms cognitively represent the fact that *there is a probability of 0.3 that the environment is in state S*. Note that, in these cases, the cognitive representation itself is still purely qualitative, but the content of that representation is graded. Therefore, on this proposal, it is entirely possible that organisms react differently to the cognitive representation that *there is a probability of 0.3 that the environment is in state S* than to the cognitive representation that *there is a probability of 0.7 that the environment is in state S*. Which kinds of cognitive representational contents it is adaptive for an organism to rely on will depend on the details of the case, but, again, the key point here is just that there is no principled problem with appealing to cognitive representational contents that concern matters of degree.

Now, the key point to note about both of these proposals is that it is plausible that they will be sufficient to make my account consistent with the appeal to graded cognitive representations in economics, psychology, behavioral ecology, etc. The main reason for this is that both of these proposals respect and accommodate the core reason for appealing to graded cognitive representations: namely, that organisms frequently seem sensitive to the strength of their evidence about the state of the world. While this sensitivity is understood in different ways on the two proposals, it is equally well respected in both cases.[15]

That said, it also needs to be acknowledged that these proposals require reading appeals to graded cognitive representations non-literally. It is not that cognitive representations themselves come in degrees of strength; rather, it is that the mechanism that causes the tokening of some cognitive representations comes in different degrees of strength, or that the content of some cognitive representations concerns matters of evidential strength. However, there is no reason to think that this is problematic: at most, this requires a reinterpretation of a commonly used modeling framework; it does not require a change in the way this framework is actually used.

All in all, therefore, the following conclusion emerges. While my account is, as such, based on purely qualitative cognitive representations, it can make sense of the fact that it is often useful to appeal to graded cognitive representations: namely, by treating these appeals non-literally.[16]

Social Learning and the Generation of Non-adaptive Cognitive Representations

There has been much work done surrounding the idea that many human— at least—cognitive representations seem to be socially acquired: as noted in chapter 3, we learn from others, and much of what we think about the world is derived from what others think about the world (Boyd & Richerson, 2005; Sterelny, 2012; Heyes, 2012; Galef, 2012; Reader & Laland, 2003). This fact may appear to have profound implications for my account: for it is widely accepted that, once cognitive representations are socially acquired, it is not only possible but in fact quite likely that many of these cognitive representations will lead to non-adaptive behavior (Boyd & Richerson, 1985, 2005).[17]

To see this, note that, in the case of social acquisition of cognitive representations, what determines whether a cognitive representation is adopted is not (just) whether that representation leads to adaptive behavior, but whether that representation has features that make its transmission and acquisition easy (Boyd & Richerson, 1985, 2005). Put differently, the fitness of a socially acquired cognitive representation is determined not by the expected reproductive success of the organisms acting on that representation, but by the expected copying success of that representation itself. As a result, cognitive representations that are easy to teach, easy to learn, or come from organisms with high status as models in the relevant population will be more likely to be acquired by others—even if they have maladaptive consequences (Boyd & Richerson, 1985, 2005).

This potential maladaptedness of the behavior resulting from many socially acquired cognitive representations is worth noting here, as one might wonder how it can be handled by my account: it appears my account presumes that the cognitive representations an organism acts on are adaptive. If that is not necessarily the case, though, then this would seem to be a cost (of sorts) that needs to be taken into account when assessing the adaptive advantages of being a cognitive representational decision maker. In other words: if cognitive representational decision makers are at a higher risk of acting maladaptively, would that not be a further reason to think that cognitive representational decision making will often *not* be adaptive?

However, as a matter of fact, the issues here are misleading. This is so for two reasons. First, it is not true that only cognitive representational decision makers can be social learners. In fact, it is well established that many organisms acquire reflexes socially as well: for example, rats seem to learn taste aversions from their parents, and something similar goes for many other animals (Galef & Laland, 2005; Mackintosh, 1994; Heyes, 2013). What is different about cognitive representational decision makers is just what they learn from others—that is, cognitive representations—and how they learn from others—for example, through language and other high-fidelity, high-bandwidth channels (see also Sterelny, 2012). Given this, though, there is no reason to think that only cognitive representational decision makers have to be concerned about acquiring possibly maladaptive behavioral dispositions from others: purely reflex-driven organisms have to do so, too. In other words, this is not a special problem about cognitive representational decision makers, and so not something that speaks specifically against the evolution of the latter.

Second, the theory of gene-culture coevolution in fact finds an easy home within my theory of the evolution of cognitive representational decision making. This can be seen most easily from noting that my account addresses the question of whether an organism should make decisions based on cognitive representations at all, whereas gene-culture co-evolutionary theory (of the relevant kind) addresses the question of which cognitive representations (or behavioral variants, or other psychological dispositions— see also Richerson & Boyd, 2005) ought to be expected to spread through a population. These are quite different questions. In fact, since the ability to act on cognitive representations needs to evolve first—before an organism can evolve the ability to acquire cognitive representations from others—the fact that many socially acquired cognitive representations lead to maladaptive actions can only be explained on the assumption that, for one reason or another, relying on cognitive representations is adaptive overall. (Indeed, this point is a key part of the project of gene-culture coevolutionary theory—see, e.g., Boyd & Richerson, 2005.) Put differently: before it is possible to ask which kinds of cognitive representations will spread in a population, it is necessary to ask why a population of organisms will rely on cognitive representations at all when making decisions. It is perfectly consistent to suggest that cognitive representations (generally) lead to adaptive behavior when answering the latter question, and yet allow for the fact that many socially acquired cognitive representations lead to maladaptive behavior.

All in all, therefore, the fact that the ability to act on cognitive representations makes an agent vulnerable to maladaptive behavioral dispositions should not be seen to speak against my account. Instead, this is something that can be easily integrated into it.

Cognitive Representational Decision Making and the Modularity of Mind

A fiercely debated issue in cognitive science and related fields for several decades has been the question of whether human and animal minds ought to be seen to be as "modular"—and if so, to what extent (see, e.g., Fodor, 1983; Tooby & Cosmides, 1992; Carruthers, 2006; Nichols & Stich, 2003; Lyons, 2001; Prinz, 2006; Samuels, 2006; Schulz, 2008; Gallistel, 1990, 2000). While there is some controversy over what exactly it means for a mind to be "modular" (see, e.g., Carruthers, 2006; Nichols & Stich, 2003; Prinz, 2006; Colombo, 2013), for present purposes, it is enough to see the thesis of the modularity of mind as referring to the idea that the mind consists of a number of different components that operate (more or less) independently of each other. For example, there might be a distinctive module responsible for determining whether another agent is cheating in social exchanges (Cosmides & Tooby, 1992), or there might a module dedicated to acquiring beliefs about which males in one's society are of high status (Buss & Schmitt, 1993). Given this, what is at stake in the debate surrounding the modularity of mind is whether there are these distinct components of human and animal minds (or at least similar ones), or whether it is better to see human and animal minds as more or less general inference and memory systems that do not have many specialized sub-components (Carruthers, 2006; Prinz, 2006; Fodor, 1983).

It is useful to consider this debate here, for my account might appear to be inherently non-modular. In particular, it might seem that, according to my account, a cognitive representational decision maker makes all decisions in the same, domain-general way: by consulting its equivalent of table 5.5 or figure 5.3. However, this is once more misleading: my account here is in fact perfectly consistent with the mind being modular (though it does not presuppose it either). More specifically, my account can allow for two different (but again mutually compatible) ways in which an organism's mind might be modularly designed.[18]

First, an organism might be disposed to acquire different cognitive representations in different ways. For example, it might operate with certain expectancies about the world, leading to the acquisition of certain specific cognitive representations given episodes of social or individual learning (see, e.g., Tooby & Cosmides, 1992; Carey, 2011; Carruthers, 2006). That is, nothing in my account precludes that the acquisition of different (sets of) cognitive representations operates in quite different ways. If this is so, though, then these different (sets of) cognitive representations can be seen as giving rise to different mental modules (Tooby & Cosmides, 1992).

Second, an organism might be disposed to build up complex cognitive representations out of simpler cognitive representations in different ways, depending on the content of these representations (Carey, 2011; Carruthers, 2006). Put differently: the patterns of inference the organism relies on might be different for different kinds of cognitive representations (social vs. non-social ones, say). If so, then these different cognitive representations can again be seen as setting up different mental modules (Tooby & Cosmides, 1992; Carey, 2011; Gallistel, 1990, 2000).

In this way, just as in the case of gene-culture coevolutionary theory, the thesis that the mind is massively modular can be easily situated *within* my account of the evolution of cognitive representations. In fact, this is also in line with some of the major recent defenses of the modularity of mind in the literature (see, e.g., Carruthers, 2006; Sperber, 2005), which also explicitly situate mental modules within a cognitive architecture spanned by (among others) cognitive representations.

All in all, therefore, it becomes clear that the account of the evolution of cognitive representational decision making defended in this chapter is consistent with a number of the major issues concerning this way of making decisions. This is important to note, as it suggests that my account does not rely on highly controversial assumptions about cognitive representational decision making.

Conclusions

I have presented and developed a new account of the evolution of cognitive representational decision making. This account is based on the idea that cognitive representational decision makers can—though need not—be more cognitively efficient than purely reflex-driven organisms: in particular, they can have an easier time adjusting to a changed environment, and they can streamline the neural machinery underlying their decision-making system. While these efficiency gains often have to be bought at the cost of slower decision making speed and more need for concentration and attention, the overall balance can, in some circumstances—such as certain complex social, spatial, and causal environments—favor the evolution of cognitive representational decision making. I have then presented some extensions of the account: in particular, I have shown that the account can be made to be consistent with the appeal to graded cognitive representations, with the fact that many human (and perhaps some non-human) socially acquired cognitive representations might lead to maladaptive behavior, and with the idea that human and non-human minds should be seen to be massively modular.

In this way, I have tried to present an account that goes some way toward fulfilling the desiderata of chapter 4: it presents some genuine benefits of cognitive representational decision makers relative to non-cognitive representational decision makers (they are more cognitively efficient), it takes seriously the fact that (cognitive) representational decision making comes with some costs (this kind of decision making will generally be slower, and requires more concentration and attention), it is relatively detailed (cognitive representational decision making can be expected to evolve in complex social, spatial, or causal environments), and it treats the evolution of cognitive and conative representational decision making separately from each other. Indeed, it is to the latter that we now turn.

6 The Evolution of Conative Representational Decision Making

What adaptive benefits can an organism receive from relying on conative representations (i.e., representations about the goals or functions of its behavior) when deciding what to do? In this chapter, I provide an answer to this question. In particular—and very much in line with chapter 5—I defend a cognitive-efficiency–based account of the evolution of conative representational decision making.[1]

The core idea behind this account is that, similarly to cognitive representational decision makers, conative representational decision makers can, in some circumstances, adjust more easily to a changed environment and streamline their neural decision-making machinery. However, as I also make clearer, the origins of these benefits are different here than in the case of cognitive representational decision making: they center on patterns in the way the organism *reacts to* the world, and not on patterns *in the states of the world* that the organism can react to. This has some important implications for the situations in which conative representational decision making is adaptive relative to when cognitive representational decision making is adaptive.

The chapter is structured as follows. To begin, I lay out my account of the evolution of conative representational decision making (taking some of the insights of chapter 5 as given). Then, I extend the account to four issues of current debate: the distinction between "ultimate" and "instrumental" conative representations, the frequent occurrence of intransitive choices, the existence of conflicts among conative representations, and the existence of non-adaptive conative representations. Next, I combine the picture laid out in this chapter with that laid out in the previous chapter to develop a clearer account of the relationship between the evolution of conative and cognitive representational decision making. Finally, I present my conclusions.

The Adaptive Benefits of Conative Representational
Decision Making

The goal of this section is to defend the idea that there can be adaptive pressures toward the evolution of conative representational decision making. These adaptive pressures stem from the fact that consulting the function or goal of an organism's behavior enables the organism to completely do away with a set of stored behavioral responses—that is, a "look-up table" that connects either perceptual cues or cognitive representations with particular behavioral responses—and simply *calculate* what to do. To understand this better, it is best to begin by briefly reconsidering some of the key aspects of conative representational decision making.

As made clearer in chapter 2, conative representational decision making is here understood as decision making that rests on representations (i.e., higher-level mental states downstream from its perceptual states) about what the organism is to do. That is, a conative representational decision maker is an organism that makes decisions by consulting the goal (understood in the behavior-generating sense laid out in chapter 2) of its behavior and, on this basis, picks the appropriate behavioral response to its environment. Two things are further important to note about this way of making decisions.

First, as noted in chapter 2, I here assume that there will be a systematic relationship between what an organism represents as being the target of its actions, and those actions themselves. That is, as also made clearer in chapter 2, I here assume that conative representations *can* play a role in the generation of an organism's behavior: the organism can decide what to do by recovering this from its goals. (This is not to say that an organism could not also represent goals that it cannot connect to its behavior; the point is just that these are not goals the evolution of which I am interested in investigating here. I return to this briefly below.)

In this way, conative representations can be seen to be akin to "behavioral rules": they state what the thing to do is for the organism. Note that, as also made clearer in chapter 2, the idea here is not that conative representations necessarily need to take the shape of imperatives ("do x"); they could also represent that a given state of the world S is rewarding, pleasurable, or something that ought to be brought out (for different views along these lines, see, e.g., Schroeder, 2004; Morillo, 1990; Damasio, 1994). What matters is just that, as I understand them here, conative representations are inherently "directive": they express, in one way or another, what the organism is to do (see also Millikan, 2002).

In turn, this leads to the second and, for present purposes, major point to note about conative representations. Given that they are inherently directive representations—explicitly represented behavioral rules of sorts—an organism can use them to infer what to do, instead of storing what it is to do in every possible situation it might find itself in, as a non-conative representational decision maker is required to do. To see this more clearly, it is best to start with a stylized example.[2]

Assume an organism has to reach a designated spot (its nest site, say) in a rugged physical terrain that does not permit the organism to perceive far from its current location and also contains no landmarks by which the organism could navigate. Assume further that this terrain can be reasonably well represented by a grid consisting of $x_{max} = 10$ and $y_{max} = 10$ points, with each point representing the center of a patch on the grid. Finally, assume, for simplicity, that the spot the organism is looking for is at or near the center of the relevant terrain—that is, assume that $x_{target} = x_{max}/2 = 5$ and $y_{target} = y_{max}/2 = 5$ (where x_{target} and y_{target} are rounded to the nearest integer). Figure 6.1 makes this clearer:

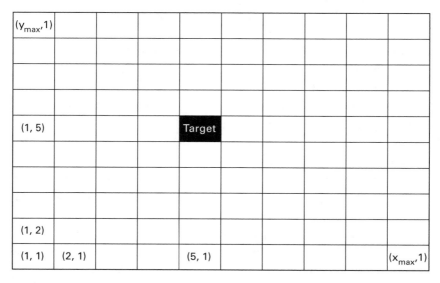

Figure 6.1
A spatial decision problem.

Table 6.1
A Non-conative Representational Decision Maker

Current Location	Behavioral Response
(1, 1)	Go north 4 patches, and east 4 patches
(2, 1)	Go north 3 patches, and east 4 patches
(3, 1)	Go north 2 patches, and east 4 patches
. . .	. . .
(1, 2)	Go north 4 patches, and east 3 patches
(1, 3)	Go north 4 patches, and east 2 patches
. . .	
(5, 5)	Stay
. . .	

Given all of this, in order to be adaptive, a non-conative representational decision maker will need to associate subsets of the terrain with different behavioral responses. Exactly how this can be done will depend on the details of the case, but for ease of exposition—and without loss of generality—I here assume that the relevant organism cannot do better than associating each patch on the grid with a given behavioral outcome. That is, I here assume that the terrain is so rugged and so difficult to navigate that there are no further patterns that the organism can pick up on in deciding how to behave, apart from focusing on what the right action is for each patch in the terrain. This is of course not a feature of all environments, but what matters is just that an adaptive non-conative representational decision maker will need to rely on something like table 6.1 to decide how to interact with its environment:[3]

Note that, in line with desideratum 1 of chapter 4, I make no assumptions about how the organism detects its location on the grid. That is, I make no assumptions about whether each position on the grid can be adaptively tracked with a few perceptual cues, or whether it is adaptive to cognitively represent that position (though I return to this issue in this chapter below). For this reason, I here leave open exactly what the left-hand column of figure 6.1 is (i.e., whether it contains cognitive representations or perceptual cues only). All that matters is that (as also made clearer in chapter 2) a non-conative representational decision maker relies on some kind of table that connects either perceptual cues or cognitive representations about

Rule: Go north $x_{target} - x_{current}$ patches, then east $y_{target} - y_{current}$ patches.

Current location: (3, 4)
Target: (5, 5)

Behavioral response: go north 2 patches, then east 1 patch.

Figure 6.2
A conative-representational decision maker.

where the organism is in the terrain with specific behavioral commands about where it is to go.

However, the situation is very different for an organism that relies on conative representations to make a decision. In particular, such an organism could simply rely on the following decision rule: "go north $x_{target}-x_{current}$ patches, then east $y_{target}-y_{current}$ patches" (where negative values imply movement in the opposite direction—that is, south or west—and $x_{current}$ and $y_{current}$ are the current positions of the organism).[4] In other words, instead of storing the right behavioral response for each spot on the grid, the organism can merely calculate—infer—what to do. All it needs to do so is information about where it is on the grid and what the goal (or function) of its behavior is. Graphically, this can be represented like in figure 6.2.

Note that this reliance on conative representations amounts to a shift away from relying on anything like the table of behavioral dispositions of table 6.1. In effect, the organism stops storing every instance of the behavioral rule or function it is acting in accordance with, and just relies on storing the rule or function itself: the organism *computes* its behavioral response to the environment by consulting an explicitly stored behavioral function and no longer relies on a stored table of all the argument-value pairs that make up this function (see also Barrett, 2005; Tooby et al., 2005). In short: the organism stops merely acting *in accordance with* a given behavioral rule, and actually *follows* that rule (for more on the distinction between acting in accordance with a rule and following a rule, see, e.g., Kripke, 1982; Davidson, 1982; Harré, 2002; Mele, 1987). Here, it is also important to note that non-conative representational decision makers can act in accordance with the same rule that conative representational decision makers do (for we may assume that the table of behavioral dispositions that underlies the former is generated from the very rule that underlies the latter)—however, they have to do it in a different, more piecemeal manner: by storing the result of every application of this rule.

Now, there are two reasons for why this shift to relying on explicitly stored behavioral functions (i.e., conative representations) can be adaptive. Interestingly, these reasons are somewhat parallel to the reasons for why relying on cognitive representations can be adaptive—though, as is made clearer below, their detailed evolutionary implications are quite different.

First, relying on conative representations can be adaptive, as it can make it much faster and cost-efficient—in terms of both cognitive resources like concentration and attention as well as energetic resources—to adjust an organism's behavior to a changed environment. To see this, return to the above example of an organism having to navigate a difficult terrain, but assume now that, for some reason, the target of the organism's navigational problem shifts—for example, perhaps its nest site was blown to a different location in the terrain.

In a case like this, a non-conative representational decision maker has to alter a (potentially vast) number of behavioral dispositions (i.e., each row in something like table 6.1): each point on the grid has to be associated with a different behavioral disposition. By contrast, a conative representational decision maker just has to change its encoding of where the target is (i.e., it has to set new values of x_{target} and y_{target}). On the—reasonable—assumption that changing each behavioral disposition in a table of behavioral dispositions takes some time, concentration, attention, and energy, the latter change will be less costly: while a conative representational decision maker also has to make a—possibly costly—change to its decision-making system, a non-conative representational decision maker has to make many more such changes.

It is important to emphasize that the size of this first adaptive benefit of conative representational decision making depends on the size of the set of stored behavioral dispositions a comparable non-conative representational decision maker would need to rely on. With small such sets, the needed changes are not very numerous, so that a conative representational decision maker might not come out on top by much (or even anything). Still, the important point is that, *in some circumstances*—that is, with extensive sets of behavioral dispositions—conative representational decision makers do come out on top.

The second reason for why relying on conative representations in making decisions can be adaptive is that it can streamline an organism's neural decision-making machinery. In particular, instead of having to store a possibly highly extensive table of behavioral dispositions, the organism just needs to store the relevant behavioral function. While the latter might require reliance on a different kind of memory—depending on whether the relevant organism already is a cognitive representational decision maker (i.e., depending on whether the left-hand side of table 6.1 consists of cognitive representations)—it is highly plausible that the neural decision-making machinery of conative representational decision makers can be more streamlined than that of non-conative representational decision makers.

To see this, note that a behavioral rule contains less information than the equivalent table of behavioral dispositions: by itself, it does not tell the organism what to do. It is just a means for the organism to derive the latter from information about the current state of the organism. Put differently: an organism that knows the state of the environment and knows what rule it should follow in this environment does not yet know what to do: it still needs to actually *apply* the rule to the state of the environment. In effect, the rule provides the organism with a means of generating something like the appropriate row in a table of behavioral dispositions—however, it does not "contain" this row (or any other one) as such. Because of this, it is plausible that relying on a conative representation—a behavioral rule—will allow the organism's decision-making system to be smaller relative to a non-conative decision-making system that is based on a set of stored behavioral dispositions: the rule still needs to be applied to a given case to yield the relevant behavioral response—this response is not already stored in the mind of the organism.

To understand this better, it is again useful to consider the situation by considering network representations of the two ways of making decisions (the conative representational one and the non-conative representational one). A network representation of (parts of) table 6.1 looks like what is shown in figure 6.3.[5]

In this network, the bottom nodes (representationally or non-representationally) detect the x coordinate of where the organism is (i.e., $x_{current}$), and the top nodes set out the relevant motor command—that is, the number of patches to the east the organism should walk (with negative values representing a westward movement).

By contrast, a network representation of figure 6.2 looks very different. Instead of simply matching input nodes to output nodes, it considers the information contained in the firing of a given node, and uses this information to *generate* the relevant output. There are many different ways in which this can be accomplished; one way to do so is by adding a central node to the network in figure 6.3 that typically fires at a given rate, but whose firing rate will be increased or decreased in different ways depending on which input node is active.[6] This can be represented as shown in figure 6.4.

In this network, the bottom nodes (representationally or non-representationally) detect the x coordinate of where the organism is (i.e., $x_{current}$). Each of these nodes then fires at a unique rate (proportional to the organism's position in the west-east dimension). The top node fires at a rate R that is computed as $R = x_{target} -$ (firing rate of input node). The organism then uses the firing rate of the top node as the relevant motor command—that is, the number of patches to the east the organism should walk (with negative values representing a westward movement).

Now, for present purposes, the key point to note is that the network in figure 6.4 is smaller—that is, it contains fewer nodes and connections—than that in figure 6.3. Intuitively, the reason for this is that is that the network in figure 6.3 is one of the least efficient ways of generating output from inputs: it connects each input node to a given output node, without working with the precise information contained in each node. By contrast, the network in figure 6.4 does exactly this: it uses the nature—that is, the content—of each input node to modulate the way the organism responds to the information it receives from the world (Hasselmo, 2012; Franco & Cannas, 1998). The exact way in which this computational modulation

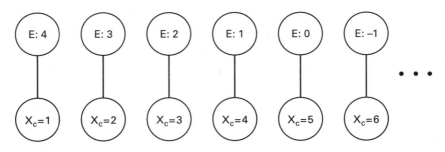

Figure 6.3
Network representation of parts of the west-east decision dimension of table 6.1.

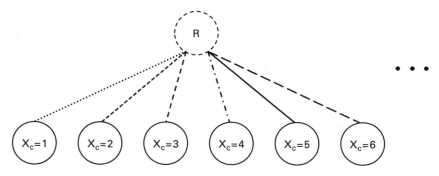

Figure 6.4
Network representation of parts of figure 6.2.

is implemented will depend on the nature of the computational system chosen (e.g., whether it is a classical or a non-classical system); fortunately, though (as also noted in chapter 2), the details of this do not matter here (see Piccinini, 2015; and Piccinini & Bahar, 2013, for more on this). All that matters here is that more computational systems can be streamlined relative to non-computational systems just in virtue of the fact that they are computational.

Now, as presented, this is a point about the *cognitive* efficiency of a conative representational decision maker. However, by the same arguments as laid out in chapter 5, there are good reasons to think that the cognitive efficiency of conative representational decision makers will be associated with greater *neural* efficiency as well (McGivern et al., 2002; Casey et al., 2000; Neubauer et al., 2005; Neubauer & Fink, 2009; Grabner et al., 2004; Rypma et al., 2006; E. Santos & Noggle, 2011; Gazzaniga et al., 2009). Specifically, it is very plausible that the neural machinery underlying conative representational decision makers can be more streamlined than that of non-conative representational decision makers: by adding a computational inference, the former can do away with a number of connections and nodes in their neural machinery (Gazzaniga et al., 2009). Put differently: by adding the resources of the neural systems underlying conative representational decision making, the organism's overall brain functioning can be made more efficient.

In turn, these savings in the neural decision-making machinery of conative representational decision makers are adaptively important—again, for much of the same reasons laid out in chapter 5. Organisms with a more efficient neural organization save valuable resources that they can spend in other ways—for example, by expanding on other cognitive or non-cognitive systems. So, a conative representational decision maker has the resources to expand on its executive control or perceptual systems, or they can increase the memory systems dedicated to storing cognitive representations of one kind of another (see also Klein et al., 2002). (This last point will become important again below.)

Now (again as in chapter 5), it is important to be very clear about the scope of this last claim. The size of these adaptive benefits of conative representational decision makers again depends on the extensiveness of the equivalent set of behavioral dispositions a non-conative representational decision maker would need to rely on. An organism that only has two behavioral dispositions—such as staying at the current patch if food is recovered at a rate greater than some fixed parameter, and leaving for a different one otherwise (see, e.g., Kacelnik, 2012)—will not gain much, if anything, by relying on conative representations. For this reason, it is important to emphasize that the claim here is not that conative representational decision making *must* come with significantly increased neural efficiency, and therefore *must* be significantly more adaptive than non-conative representational decision making. The claim is just that it will come with significantly increased neural efficiency *in some circumstances*—namely, those that feature an extensive set of behavioral dispositions that can be replaced with an easily computable function—and that it will therefore be more adaptive in *those* circumstances.

In short: conative representational decision makers can, in some cases, increase the ease with which they adjust to changed environments and streamline their neural decision-making machinery (relative to non-conative representational decision makers). In turn, this means that there are some circumstances where the reliance on conative representations when making decisions is adaptive. In line with the arguments of chapter 3, this thus provides *a reason* (nothing more, but also nothing less) to expect—in the relevant environments—the evolution of conative representational decision making.

However, this is once again only half of the story—for, as noted in chapters 2, 4 and 5, relying on mental representations also comes with costs that need to be taken into account. However, since these costs are the same as the ones considered in the previous chapter concerning the evolution of cognitive representational decision making, I can be briefer here.

In particular, as noted earlier, the major costs that conative representational decision makers face are, first, the fact that it is likely conative representational decision making is slower than non-conative representational decision making, and, second, the fact that conative representational decision making is likely to require more in terms of concentration and attention than non-conative representational decision making. Both of these types of costs come out very clearly from noting that conative representational decision makers, by their very nature, have to *infer* what they are to do. This need for making inferences is likely to slow down the organism and also to take up concentration and attention that it could use for other purposes.[7]

What this implies is that, once again, conative representational decision making should only be expected to evolve when the benefits that it brings to an organism—the increased ease in adjusting to changed environmental circumstances and the more streamlined neural machinery underlying decision making—outweigh the losses it brings to an organism—the increased demands for decision making time, concentration, and attention. Put differently: conative representational decision making is adaptive if it is more efficient *overall* than non-conative representational decision making—that is, when its efficiency losses *at a time* are outweighed by its efficiency gains *across time*.

This means, more concretely, that conative representational decision making is adaptive—in net terms—when (a) non-conative representational decision making would need to rely on a very extensive set of behavioral dispositions, and (b) when the behavioral function that the conative representational decision maker would rely on is relatively easy to compute. Point (a) holds, since with extensive sets of behavioral dispositions, the adaptive benefits of conative representational decision making are high (as noted earlier). Point (b) holds, as with easily computable behavioral functions—that is, with decision rules that it is easy to apply to a given case—the costs of conative representational decision making are likely to be quite low. There are two further considerations that it is important to note concerning this argument.

First (and similar to what is true when it comes to cognitive representational decision making), what it means to be "easily computable" will depend on the specifics of the organism in question. Organisms with relatively good inferential powers can make relatively more complex inferences relatively more easily. Of course, the inferential abilities of organisms can also evolve; the point that is relevant here is just that, at any given time, the inferential abilities of an organism—as it is then—determine which behavioral functions it can easily compute. In turn, this implies that there is no general standard holding for all organisms at all times that determines which conative representations they can be expected to be able to rely on. However, this is consistent with the possibility of "local" comparisons: it can be possible to determine whether, relative to the cognitive abilities of a given organism at a given time, a given conative representation is likely to be too complex to evolve. For example, it is known that, despite its apparent adaptiveness, reciprocal altruism is rare in nature (see, e.g., Hammerstein, 2003; Noe & Voelkl, 2013). A reason commonly given to explain this is that, for many organisms, keeping track of who helped who in the past is quite difficult (see, e.g., Hammerstein, 2003; Noe & Voelkl, 2013). If that is so, then it is implausible to assume that many organisms can rely on conative representations that require them to track these sorts of facts. (I return to this issue briefly in chapter 9.)

The second point to note here is that it is important to realize there are often several different behavioral functions, embodying different levels of computational complexity, that an organism could rely on to replace a given table of behavioral dispositions. This point will become important again in chapters 7 and 8 below, but for now, it is sufficient to note that, ceteris paribus, we should expect the evolution of the simplest—that is, the most easily computable—available behavioral rule.

In a nutshell: we should expect the evolution of conative representational decision making in cases where this yields large benefits relative to its costs. Now, in line with the third desideratum of chapter 3, it is desirable to state more specifically which kinds of environments are characterized by this abstract description. Fortunately, just as in chapter 5, it is possible to sketch the outlines of three of the major kinds of environments that make conative representational decision making adaptive. Furthermore, just as in chapter 5, these environments comprise certain cases of spatial navigation, certain cases of causal navigation, and certain cases of social navigation. However, while the types of environments that make conative representational decision making adaptive thus overlap with those that make cognitive representational decision making adaptive (a point to which I return below), the *reasons* for why these environments make conative representational decision making adaptive are slightly different from the reasons for why they make cognitive representational decision making adaptive. To see this, consider these three types of environments in more detail.

First, as made clearer in the above stylized example, a number of organisms face problems of spatial navigation that require a large number of behavioral dispositions to handle, or which can be successfully solved using a relatively easily computable behavioral function. Cases like this include the problems of some types of insects to find flowers in relatively homogenous meadows, or the problems of a number of animals such as prairie dogs, ferrets, meerkats, or prairie voles that have to return to their burrows after going out to forage (see, e.g., Phelps & Ophir, 2009).

Concretely, consider elephant shrews (for more on the ecology of these animals, see, e.g., Rathbun & Redford, 1981; Linn et al., 2007; Fitzgibbon, 1995). Some species of these animals make tracks throughout their hunting environment, which they then use both to hunt for food and to escape from predators. In order for this to work, though—that is, in order for an elephant shrew that is out hunting insects to successfully escape from a predatory lizard—the animal needs to be able to find its way quickly from wherever it is in the environment back to its underground burrow. Importantly, navigation by landmarks is, due to the homogenous nature of the environment, generally quite difficult. Hence, this sort of case exemplifies precisely the sort of scenario where conative representational decision making is adaptive.

Second, there are also some "causal" (in a broad sense) environments in which conative representational decision making will be adaptive. In particular, many (though not all) foraging problems are likely to be of this kind. So, for example, organisms that track, hunt, or forage for many different kinds of prey or foodstuffs, each of which requires a different application of the same overall foraging strategy to be successful, can be assumed to find conative representational decision making to be adaptive.

Concretely, consider chimpanzee (or early human) hunting practices. Here, it is plausible that the adaptiveness of hunting a given prey animal will depend on complex combinations of the size of the hunting party (larger groups make it more likely that the hunt will be successful, but also entail less food reward per participant) and the size of the prey animal (larger prey animals yield more food reward per hunting participant, but are also harder to hunt successfully; see also Boesch, 1994, 2002; Stanford, 1995; Mitani & Watts, 2001; Hawkes & Bird, 2002; Sterelny, 2012; Mithen, 1990).

In a case like this, a non-conative representational decision maker would need to store many different combinations of hunting party size and prey animal size to decide whether to join a hunt or not. By contrast, a conative representational decision maker can simply calculate some function as "join a hunt for which the ratio [size of the prey animal (in pounds) / size of hunting group] is maximized, as long as 0.5 < [size of the prey animal (in pounds) / size of hunting group] < 1.5."[8] That is, they can decide whether to join a hunt—and if there are several possible options, which hunt to join— by assessing if the spoils per person would be less than half a pound of prey animal (in which case hunting would waste too much energy relative to what it can yield) or if they would be more that 1.5 pounds per person (in which case the hunt would be unlikely to be successful). Given the above, it seems plausible that the conative representational way of making decisions is more adaptive than the non-conative representational one.[9]

The third kind of environment that is likely to make conative representational decision making adaptive are some social environments. In particular, these will be environments where an organism needs to react to many different combinations of social relations, each of which instantiates a relatively straightforward pattern. So, for example, assume that whether it is adaptive to feed from a given food source depends on the presence of higher-ranking group members and the relationship between the organism acting and other group members in the situation (see, e.g., Mitani & Watts, 2001; Boesch, 2002; Hawkes & Bird, 2002). Just as in the case of foraging, a non-conative representational decision maker would need to store many different combinations of the latter two relations to make adaptive feeding decisions. By contrast, a conative representational decision maker may be able to just rely on the following behavioral function: "feed if $R \times D \leq 1$," where R is some measure of the relatedness of the organism to others present (say, with $R = 0$ for parents, $R = 1/2$ for siblings, and $R = 1$ for everyone else), and D is the number of ranks by which others present outrank the acting organism.[10] Given the above, conative representational decision making is again likely to be more adaptive than non-conative representational decision making here.[11]

In short: there are a number of concrete environments that fit the description of situations that favor conative representational decision making. An important final point worth noting about these environments is that, similar to chapter 5, they share important parallels with (though they also differ in some important ways from) the cases identified by other researchers concerning when and where representational decision making is adaptive. In particular, like Millikan (2002), Papineau (2003), Dickinson and Balleine (2000), Sterelny (2003), and Koscik and Tranel (2012), I also identify relatively complex spatial, causal, and social environments as crucial for when conative representational decision making is likely to be adaptive. However, as compared to these other researchers, I identify different features of these environments as being especially important for the adaptiveness of conative representational decision making. In particular, according to my account, the *nature* of complexity of the relevant environments matters greatly for whether they favor the evolution of conative representational decision making: while it is true that the environments need to be complex on a surface level, on my account, they also need to be relatively simple on a more fundamental level. More specifically, there needs to be a relatively easily computable function that underlies the environmental complexity, for otherwise the costs of conative representational decision making are likely to outweigh its benefits.

Putting all of this together, this yields the following picture. A key reason speaking in favor of the evolution of conative representational decision making is the fact that, by relying on an inference from an explicitly represented goal to a specific behavior, this way of making decisions can help organisms easily and quickly adjust to changed environments and streamline their neural decision-making machinery. While conative representational decision making—like all representational decision making—also comes with costs, there are circumstances in which the benefits are likely to outweigh the costs. These circumstances include specific kinds of complex spatial, causal, and social environments: namely, situations in which the organism has to react to a large number of different circumstances, but where each of these reactions is an instantiation of a relatively simple pattern.

Extensions

I now consider four extensions of the account laid out in the previous section. As was the case in chapter 5, the goal in this is twofold: first and most importantly, it is to show how my account can incorporate various questions and debates that concern conative representational decision making more generally. Second, the goal is to make the account clearer by placing it in the context of these other questions and debates. The four extensions I consider concern the differences between instrumental and ultimate conative representations, the frequent occurrence of intransitive choices, the existence of conflicting conative representations, and the existence of non-adaptive conative representations.

Instrumental versus Ultimate Conative Representations

A distinction that is frequently made in the literature on conative representational decision making is one between instrumental and ultimate conative representations—or, as it is more commonly put, between ultimate and instrumental "desires" or "preferences" (see, e.g., Alvin Goldman, 1970; Sober & Wilson, 1998; Stich, 2007; Nichols & Stich, 2003; Schroeder, 2004). While it is controversial exactly how to best draw this distinction (Alvin Goldman, 1970; Stich, 2007), for present purposes, it is enough to note that this distinction is widely seen to be based on the origin of the relevant conative representations. In particular, a conative representation is said to be "ultimate" if it is not derived—in some sense of this term—from other conative representations, and it is said to be "instrumental" if it is so derived. So, an organism that has an ultimate conative representation to achieve a particular kind of goal—to eat a particular food stuff, say—differs from one that has instrumental conative representation to achieve that particular kind of goal, in that the latter's conative representation has been derived from some other conative representation—say, for eating in general (see also Schroeder, 2004).

Now, on the face of it, my account of the evolution of conative representations seems to be merely an account of ultimate conative representations. In particular, my account is based on the idea that organisms that are conative representational decision makers reason *from* a given conative representation *to* a given form of behavior. However, nothing in my account seems to straightforwardly correspond to instrumental conative representations—for nothing in my account seems to depend on (or make reference to) the inference *toward* a conative representation. This might seem to be problematic, in that it would seem to make it hard to make sense of the frequent appeal to instrumental conative representations made in the literature. However, for two reasons, this is not nearly as problematic as it might first seem.

First, it is in fact possible to allow for some instrumental conative representations on my account. In particular, it is entirely possible that an organism, in reasoning from a given behavioral function—that is, an "ultimate" conative representation—generates various sub-goals—that is, "instrumental" conative representations—which it then, in turn, uses to calculate the behavioral response appropriate to its environment. The main reason for why an organism might find the generation of these kinds of sub-goals useful is that it enables it to circumvent the constraints that are invariably part of actual cognitive systems. A good example of this is the fact that humans—and, plausibly, most other cognitive systems as well—can only hold a small number of items in their working memory at the same time (Kareev, 2012), so that, in lengthy computations, it becomes useful (and in fact necessary) to divide the problem into several stages, each of which is solved successively (Newell et al., 1958; J. R. Anderson, 1993; Korf, 1987; see also Bratman, 1987).[12]

The second reason for why my account should not be seen to have a major problem with the distinction between "ultimate" and "instrumental" conative representations: it is not clear that this distinction needs to always be seen to correspond to psychological reality. In particular, some researchers (see especially Sterelny, 2003, pp. 87–88) have argued that, frequently, there is no good reason to posit genuine instrumental conative representations: all that an organism needs is a densely textured set of cognitive representations about how to achieve the relevant behavioral goal. Note that this of course is not to say that there is no use talking *as if* an organism formed an instrumental conative representation. Indeed, describing a given decision-making episode in terms of the generation of instrumental conative representations may make it clearer exactly how the organism arrived at the behavioral response it chose. However, this is not to be confused with seeing these descriptions as psychologically accurate. Put differently: many of the references to instrumental conative representations made in the literature might be better seen to be elliptical statements that can be spelled out in a way that does not involve reference to these kinds of conative representations.

In short: the frequent appeal to a distinction between "ultimate" and "instrumental" conative representations does not pose problems for my account of the evolution of conative representational decision making. To the extent that there is some psychological reality to this distinction, it can be handled by my account—namely, by the fact that the generation of (temporary) behavioral sub-goals can be computationally plausible—and to the extent that there is no psychological reality to this distinction, there is nothing here for my account to handle.

Intransitive Choices

There is much recent work—both concerning human and non-human animals—that appears to show many organisms frequently make decisions that are intransitive (Broome, 1991, 1999; Guala, 2005, pp. 98–105; A. I. Houston et al., 2007; Johnson & Busemeyer, 2005; Rieskamp et al., 2006; Sopher & Gigliotti, 1993; Tsai & Bockenholt, 2006; Waite, 2001). That is, it seems that many organisms choose option A in a choice between A and B, option B in a choice between B and C, and yet option C in a choice between A and C. This is puzzling, in that the first two choices would seem to imply that the organism should choose A over C—after all, it appears that if, according to the organism's behavioral function, A is more "choiceworthy" than B, and B is more "choiceworthy" than C, then C should also be more "choiceworthy" than A.

This point is important to address in this context, in that it may appear as if my account of the evolution of conative representations implies that an organism's choices should be expected to be transitive. For example, return to the above case of the organism that has to decide whether to join a given hunting party. Here, it may happen that an organism chooses joining hunt group A over not hunting at all, as 0.5 < [size of the prey animal (in pounds) / size of hunting group A] < 1.5. Further, it may happen that, in a choice between two possible hunting parties A and B, the organism chooses B over A, as 0.5 < [size of the prey animal (in pounds) / size of hunting group B] < 1.5, but also [size of the prey animal (in pounds) / size of hunting group B] > [size of the prey animal (in pounds) / size of hunting group A]. Given this, it is clear it would then also choose joining hunting group B over not hunting at all, as, by assumption 0.5 < [size of the prey animal (in pounds) / size of hunting group B] < 1.5. Put differently: it seems that the choice here should definitely be transitive.

If this holds generally, though, how can my account make sense of the fact that many organisms often appear to choose intransitively? In fact, there are two main ways in which it can do so.[13]

First and most importantly, there is no reason to think that every organism needs to rely on behavioral functions that are "monotonic," as in the above example of the problem of which hunting party to join. For example, it is also possible that an organism relies on a behavioral function of the following sort: "join any hunt for which 0.5 < [size of the prey animal (in pounds) / size of hunting group] < 1.5; in case of several options, choose the first one offered." In a case like this, it can be the case that, in a choice between three hunting parties, each of which has 0.5 < [size of the prey animal (in pounds) / size of hunting group] < 1.5, an organism chooses to join party A over B (as A was offered first), party B over C (as B was offered first), and C over A (as C was offered first). In other words, there is nothing inherent in relying on behavioral functions that requires choices to be transitive.

Of course, this then raises the question of when an organism would rely on decision rules that are monotonic, and when it would rely on non-monotonic ones. Now, the answer to this question will depend on the details of the relevant case: how complex the relevant decision rules are (and thus, how difficult it is to compute them), and how problematic and frequent intransitive choice behavior would be. I return to issues related to this question in chapter 8; all that matters for now is that, as such, my account is not committed to monotonic choice rules—and thus, to transitive choice behavior.

Second, even in the case of monotonic choice rules, an organism can end up choosing intransitively. The reason for this lies in the way reliance on the relevant choice rules is instrumented. In particular, since applying a behavioral function to a given set of circumstances typically requires estimating various quantities, it is possible—and in fact highly plausible—that this estimation will often be less than fully precise. Indeed, there are some good reasons to think that many such estimates are made *relationally*: for example, there is good evidence that humans and many other organisms assess heights and other physical magnitudes comparatively—that is, estimate the height of a given object by comparing it to some other object (see, e.g., Krantz, 1972; Shepard, 1981). If that is so, though, then there can be intransitive choices even with monotonic decision making: for, in applying the rule to different circumstances, the organism might estimate the relevant variables in different ways, leading to intransitive choices.

To see this, return to the above foraging problem, but assume that the organism estimates the size of a given hunting party by comparing it to some reference group. In a case like this, an organism might choose hunting with party B over hunting with party A, since by comparing A with B, B is clearly estimated to be the smaller party, but still one that satisfies 0.5 < [size of the prey animal (in pounds) / size of hunting group] < 1.5. Further, the organism might choose hunting with party A over not hunting at all, as the size of A is estimated by comparing it to the size of the last hunting party the organism joined, and it then turns out that 0.5 < [size of the prey animal (in pounds) / size of hunting group] < 1.5. However, it may then still be the case that the organism chooses not hunting over hunting with party B, as the size of B is now estimated by comparing it to the last hunting party—which may be different from the one that was the comparative base in the previous situation—the organism joined, but given this, [size of the prey animal (in pounds) / size of hunting group] < 0.5. In this way, there can be intransitive choice even with monotonic behavioral functions.

All in all, therefore: my account of the evolution of conative representational decision making does not have any problems with accounting for intransitive choices. In fact, given the fact that it is psychologically plausible that many organisms assess various magnitudes comparatively, and given the fact that it may well be plausible that there are situations where reliance on non-monotonic choice rules is adaptive, the account even *predicts* that many organisms will often make intransitive choices. This speaks in favor of the account, since, as noted above, this is exactly what we do find.[14]

Conflict among Conative Representations

A third issue worth discussing here—if only briefly—concerns the question of how an organism is to adjudicate conflicts among different conative representations. As will also be made clearer in chapters 8 and 9, it is highly plausible that most conative representational decision makers have more than one "ultimate" conative representation—that is, more than one behavioral function that drives their behavior. Furthermore, it is also plausible that these different conative representations will often conflict: the organism cannot engage in behavior that fulfills all of these conative representations equally well. Because of this, the organism needs some way of deciding which of these conative representations to make the basis of its behavior.[15]

Fortunately, my account of the evolution of conative representational decision making is consistent with many different ways of resolving such conflicts among different behavioral functions. In particular, there are two main accounts of how the coordination among different conative representations could be achieved. Note that these are not mutually exclusive, and an organism might rely on both of these in different contexts.

First, the organism could give more or less strict priority to one conative representation over another: that is, each conative representation could be assigned a given weight, and the decision as to which conative representation to make the basis of its behavior can then be determined by these weights. For example, the organism could always prioritize its conative representation of obtaining matings with members of the opposite sex, and switch to prioritizing its conative representations of obtaining food only when food reserves fall below a given threshold level (see also Laland & van Bergen, 2003, p. 163; Glimcher et al., 2005; Carruthers, 2006, chap. 2; Sterelny, 2003, chap. 5).

Second, the organism could take into account (what it takes to be) the consequences of acting on the different conative representations in question. So, instead, of strictly prioritizing one conative representation over another, the organism might apply each (or at least a subset of) the relevant conative representations to the case at hand, but do so in an "offline" manner. That is, instead of actually *acting* on the relevant behavior, the organism could "imagine" (of sorts) what acting on the given conative representation would be like: for example, it could let the conative representation generate not a behavior, but a cognitive representation of the consequences that (likely) would result from this behavior. It could then decide which of these "imagined" behaviors to actually put into practice by assessing which has consequences with a higher value. Note that the notion of "value" at stake here is very thin: the idea is just that organisms do not give weights to conative representations as such, but just to (real or hypothetical) behavioral outcomes. What constitutes these weights—whether affective states, reward states, or something else entirely—is a different question that does not affect the core of this picture (for accounts of this sort, see, e.g., Carruthers, 2006, chaps. 2 and 6; Schroeder, 2004).

Two points are important to note about these two proposals for adjudicating conflicts among conative representations. On the one hand, reliance on them has different benefits and drawbacks for an organism: the view that assigns weights to conative representations themselves is arguably easier to implement, and most likely significantly faster and less needy in terms of concentration and attention as compared to the view that assigns weights only to the behavioral outcomes of different conative representations. By contrast, implementing the latter view can be more accurate, as simple dominance relations among behavioral goals often do not exist.

On the other hand, both of these views are perfectly consistent with the account of the evolution of conative representational decision making laid out here. Put differently: my account does not depend on either of these views turning out to be true, nor does it favor one of them over the other. The reason for this is that my account—unlike that of Sterelny

(2003)—locates the adaptive benefits of conative representational decision making in the nature of that kind of decision making itself, not in the way that kind of decision making can resolve conflicts among the different behavioral goals of an organism.

Non-adaptive Conative Representations

There is no doubt that humans—at the very least—have non-adaptive behavioral goals. From aiming to be professionally or artistically success- ful even if leads to fewer children and a shorter lifespan to aiming to live up to religious convictions that forbid reproduction, humans clearly make decisions based on non-adaptive conative representations. Given that my account of the evolution of conative representational decision making is selection-based, this may thus appear to create some conflict: if conative rep- resentational decision making is adaptive, how can it lead to non-adaptive conative representations?

However, just as in the case of cognitive representational decision making, there is in fact no conflict here. In particular, my account is consistent with all of the major accounts of the sources of conative representations: these representations can be innate, individually learned, or socially acquired in many different ways (see chapter 5 for more on this). For this reason, all of the issues that arise when it comes to these sources can also be incorporated into my account here. For example, if conative representations are sometimes acquired socially, then there is much potential for non-adaptive conative rep- resentations to underlie an organism's behavior: for, then, which representa- tions are adopted may depend as much on which representations are easy to learn as on how adaptive they are (see also Boyd & Richerson, 2005).

In general, my claim is just that conative representational decision making can be more adaptive than non-conative representational decision making, since it enables an organism to achieve its behavioral goals more efficiently. However, this is as such perfectly consistent with these behav- ioral goals themselves being non-adaptive. Of course, in the long run, one would expect organisms whose behavioral goals are all non-adaptive to go extinct; however, this is as true for non-conative representational decision makers as for conative representational decision makers. Moreover, it is entirely possible for organisms to thrive even if many of their behavioral goals are maladaptive (Boyd & Richerson, 2005).

All in all: the existence of non-adaptive conative representations does not pose a problem for my account of the evolution of conative representational decision making either. There are many different explanations available for the existence of these kinds of conative representations, and, in general, my account is not focused on the *content* of the relevant conative representations, but on the mere fact that they are conative representations.[16]

The Evolution of Conative and Cognitive Representational Decision Making

It is now time to put together the account laid out in this chapter with the one laid out in the previous chapter. In particular, it is now time to consider the relationship between the evolution of conative *and* cognitive representational decision making. The reason why it is useful to do this is that, as made clearer in chapter 4, it is not plausible to think that the evolution of conative representational decision making is conceptually linked to the evolution of cognitive representational decision making. Given this, the relationship between the evolution of these two traits is not obvious, and needs to be considered separately.

The first point to note in this context is that the environments in which conative and cognitive representational decision making are adaptive are overlapping—but not fully so. In particular, for both conative and cognitive representational decision making to be adaptive, the following two conditions need to hold.

(a) In order to act adaptively, the organism needs to robustly track the relevant states of the world, and this tracking can be relatively easily and quickly accomplished in an inferential manner (i.e., by constructing the occurrence of a given state of the world out of the occurrence of simpler states of the world).

(b) The organism's behavioral reactions to the relevant states of the world can be easily and quickly (for the organism in question) computed from a behavioral function.

I think there are several more concrete environments that might satisfy both of these conditions. For example, several foraging situations might be of this kind: in particular, many foraging problems require tracking complex states of the world—where a herd is moving to, at what speed, etc.—across an extended period of time, and can be solved by using relatively straightforward behavioral rules, such as the one mentioned above concerning the hunting behavior of chimpanzees or early humans (see also Sterelny, 2003, 2012). The same goes for various social environments: living in a chimpanzee troop, for example, appears to require tracking the different hierarchical relationships in the group (and possibly also the mental states of other members of the troop—though see also Sterelny, 2003, chap. 4), and then taking all of this information into account in a rather straightforward manner to make adaptive decisions.

For this reason, it is plausible that, frequently, organisms will find both conative and cognitive representational decision making to be adaptive. This thus suggests—in the evidential way laid out in chapter 3—that many organisms in fact have evolved both conative and cognitive representational decision-making systems. In particular, at least a number of mammals— including the great apes—quite clearly seem to have been inhabiting environments (and that for significant periods of time) that favor both representational tracking of the world as well as relying on representations of what the organism is to do (see, e.g., Whiten & Byrne, 1997; Humphrey, 1986; Sterelny, 2003, 2012).

However, importantly, this need not be true for all organisms at all times. In particular, it is not necessarily the case that *all* environments that favor conative representational decision making also favor cognitive representational decision making—and the reverse.[17] The major reason for this is that these two ways of making decisions respond to slightly different features of their environment. Cognitive representational decision making relies on there being patterns in the environment that are worth picking up on, whereas conative representational decision relies on there being patterns in the organism's behavioral responses to the environment that are worth picking up on. Put differently: while both cognitive and conative representational decision makers streamline their decision-making machinery, they streamline it different ways: either by adding higher-level mental representations that group perceptual states (for cognitive representational decision makers), or by adding higher-level mental representations that transform perceptual states into behavioral commands (for conative representational decision makers). Importantly, the adaptiveness of these two can come apart.

In particular, it may be that an organism can track the relevant features of its environment using a handful of simple perceptual cues that do not need to be combined in complex ways, but it may still be adaptive for the organism to *compute* its response to the presence of the relevant perceptual cues. So, as in the example of chapter 2, an organism might make its behavior dependent, in a rule-governed manner, on purely perceptually detected differences in the wavelengths reflected off of different objects. This may be because tracking the wavelengths/colors is easily and reliably done with simple perceptual cues, but the organism's reactions to these perceptual cues is based on a sufficiently rich and easily computable function of these wavelengths (e.g., "explore the object for s seconds, with s = wavelength in nm of the object") to make reliance on an explicit encoding of this function adaptive.

Similarly, the reverse is possible as well: for example, as in chapter 5, an organism might find it adaptive to react to whether it has been sunny for two days, it has been raining for two days, or it has been alternately raining and sunny—that is, to form cognitive representations—but it might still be adaptive for it to rely on something like a table of behavioral dispositions (rather than an explicitly stored rule like "if the weather is variable across two days, forage, otherwise relocate") to manage its behavioral reactions to these different states of the world. This may be because the organism, in order to behave adaptively, needs to rely on many perceptual cues to track the weather on a given day, but once it has done so, the combination of these different cognitive representations only involves fairly few options (rain-rain, rain-sun, sun-rain, sun-sun) that can be more quickly and easily handled by not relying on an explicitly represented rule.[18]

In a nutshell: while it is plausible that many organisms find both conative and cognitive representational decision making adaptive, this is not necessarily the case in all circumstances. Moreover, even if there are reasons to think that an organism might find both kinds of representational decision making to be adaptive, there can be long periods where an organism will rely on only one of the two. On the one hand, its environments may only recently have changed to make both kinds of representational decision making adaptive. On the other hand, since the two traits evolve independently of each other, there can be lags between the evolution of one and the evolution of the other: even if both T and T' are adaptive in a given environment, they do not need to evolve at the same time if they evolve independently of each other (see also Schulz, 2013d). Two further points are important to note concerning the evolution of *both* conative and cognitive representational decision making.

First, it is plausible that the evolution of one of the two representational decision-making systems makes the evolution of the second one slightly more likely than it would be by itself. This is so for two reasons. On the one hand, it is plausible that acquiring the ability to form mental representations at all (whether cognitive or conative) might require some relatively major changes to an organism's cognitive system: for example, the organism might need to acquire a semantic form of memory. In turn, this need for some complex changes in an organism's cognitive architecture implies that, once one of the two representational decision-making systems has evolved, the second one is more likely to do so, too: after all, the "acquisition costs" of this second system have been lowered relative to the first one since much of the necessary machinery—such as a semantic form of memory—is already present in the organism. On the other hand, the fact that an organism already is a representational decision maker (of one kind or another) can make adding a second representational decision-making system more adaptive. In particular, organisms who are already cognitive representational decision makers might have an easier time picking up on patterns in their responses to the world as compared to purely reflex-driven organisms—and the reverse.

The second point worth noting about the evolution of both conative and cognitive representational decision making concerns the order in which the two systems evolve. Now, in principle, both logically possible sequences are also empirically possible: it could be that a population of organisms switches from mostly containing purely reflex-driven decision makers to containing mostly cognitive representational decision makers, and then evolves to mostly contain full (i.e., cognitive and conative) representational decision makers. Equally, it could be that a population of organisms switches from mostly containing purely reflex-driven decision makers to mostly containing conative representational decision makers, and then evolves to contain mostly full (i.e., conative and cognitive) representational decision makers. The first case would see the evolution of full representational decision making as proceeding by first streamlining the organism's table of behavioral dispositions, and then, second, by replacing the entire table with a behavioral function. The second case would see the evolution of full representational decision making as proceeding by first replacing the organism's table of reflexes with a behavioral rule, and then streamlining this rule by having it range over cognitive representations instead of perceptual cues.

However, while both of these options are principled possibilities, there are some empirical reasons to think that, at least in mammals, the second option is more plausible than the first. This is because, at least in mammals, a number of the parts of the brain that appear to underlie conative representational decision making—such as the reward system and related cortical regions—are significantly older than the parts of the brain underlying cognitive representational cognition (see, e.g., Schroeder et al., 2010, pp. 79–83 and 103–105).[19] In turn, what this suggests is that, in mammals at least, conative representational decision making evolved first, and cognitive representational decision making evolved afterward.

That said, this conclusion needs to be seen to be somewhat restricted, in that the neural systems of other organisms are generally not as well understood as those of mammals.[20] For example, we do not know much about the evolutionary history of the different parts of insect, spider, or avian brains (see, e.g., Reader & Laland, 2003). Because of this, it is not clear to what extent the situation in mammals generalizes to other animals: it is possible that, in other orders of animals, cognitive representational decision making evolved before conative representational decision making. Still, it is noteworthy that, at least in mammals, conative representational decision making seems to have evolved first: it at least *suggests* that conative representational decision making may be slightly easier to evolve than cognitive representational decision making. In turn, this ease may be due to the fact that the adaptive benefits of conative representational decision making are slightly stronger and more widely instantiated than those of cognitive representational decision making. However, further confirmation of this point awaits the further investigation of the evolutionary history of non-mammalian brains.

Summing all of this up: there is good reason to think that some—though not all—animals are both conative and cognitive representational decision makers. In addition, there is good reason to think that conative representational decision making evolved first, and that cognitive representational decision making evolved second, at least in a large group of animals.

Conclusions

I have presented and defended an account of the evolution of conative representational decision making that is consistent with the desiderata of chapter 4: it pays attention to the benefits as well as the costs of representational decision making, it is relatively detailed in predicting the environments in which conative representational decision making should be expected to evolve, it does justice to the abilities of non-conative representational decision makers, and it treats the evolution of conative representational decision making separately from that of cognitive representational decision making.

In particular, I have identified the fact that conative representational decision makers can infer what to do in a given situation, rather than needing to store this, as a major adaptive benefit of this way of making decisions: it enables organisms to adjust quickly and easily to a changed environment and to streamline its neural decision-making machinery. While this benefit needs to be compared to the costs of making decisions using conative representations—the need for more decision making time, attention, and concentration—it is plausible that there are some circumstances where this balance is positive. Specifically, conative representational decision making is likely to be adaptive in certain kinds of complex spatial, causal, or social environments: namely, in environments where organisms have to react in different ways to many different circumstances, but where each of these reactions instantiates a relatively simple pattern.

I have then shown how this account can handle a number of further issues: a distinction between ultimate and instrumental conative representations, the frequent occurrence of intransitive choices, the existence of conflicts among different conative representations, and the existence of non-adaptive conative representations. In this way, I hope to have made clear how my account can be extended to cover many of the key issues surrounding conative representational decision making.

Finally, I have put together this account of the evolution of conative representational decision making with the account of the evolution of cognitive representational decision making laid out in the previous chapter. Here, I have argued that (a) some—though not all—organisms are indeed likely to have evolved both conative and cognitive representational decision-making systems, and also that (b) there is some reason to think that, at least in one large group of organisms—the mammals— conative representational decision making evolved first, and cognitive representational decision making second.

An overarching feature of this account of the evolution of representational decision making is worth emphasizing again here: this account does not locate the benefits of representational decision making in the generation of behavioral abilities that are unavailable to non-representationally driven organisms; rather, it locates them in the fact that, at times, representationally driven organisms simply make decisions more efficiently than non-representationally driven organisms. Of course, once representational decision making is in place, it can, over time, lead to abilities that are difficult to emulate purely with reflexes (such as the generation of works of art or science). The point is just that, at least initially, this is not what drove the evolution of representational decision making. This is important to note, as it differs quite strongly from the other accounts of the evolution of representational decision making sketched in chapter 4: these all see representational decision makers as being able to do things that non-representational decision makers cannot, and not merely as being able to do the same things *better* (i.e., more efficiently).

In this way, the core picture of the evolution of representational decision making to be defended in this book is complete. In what follows, I apply this picture to a number of open questions in psychology, social science, and philosophy. To do this, I focus the discussion on organisms that have evolved both cognitive and conative representational decision-making systems. (A number of the conclusions reached also apply to merely cognitive representational or merely conative representational decision makers, though I will generally not make this explicit in what follows.) With this in mind, consider the first application of the account of the evolution of representational decision making here defended: the question of whether and when representational decision making should be seen to extend beyond the skin of an organism and into its environment.

III Applications

7 Extended Representational Decision Making—A Synthesis

Does representational decision making just happen "in the head," or does it also extend—at least sometimes and at least partially—into the environment? This question is noteworthy not just for its inherent interest, but also for being an instantiation of a more general issue that is currently much discussed in (parts of) cognitive science, psychology, philosophy, and biology: namely, the question of whether and how much of cognition in general extends into and is embedded in the environment. In particular, for some time now, a number of researchers have been arguing that many aspects of cognition are not limited to the brain (or even to other parts of the organism's body), but also involve, in a particularly deep sense, parts of the environment: much of cognition is said to be so deeply interwoven with the environment that it literally takes place, at least partly, outside of the skin of the organism (see, e.g., Clark, 1997; 2008; Rowlands, 2010; Shapiro, 2004; Stotz, 2010; Griffiths & Stotz, 2000; Schulz, 2013c; but see also Adams & Aizawa, 2008, and Rupert, 2009; for an early forerunner of this sort of approach, see Gibson, 1966).[1]

What makes considering this issue especially important in the context of the present book is that defenders of the extendedness of cognition often contrast their view with a representationalist view of cognition (see, e.g., Brooks, 1991; Varela et al., 1991; Beer, 1990; van Gelder, 1995; Di Paolo, 2005; Thompson, 2007; and see also Clark, 2008—though Clark's work is generally somewhat ambiguous on this point). In particular, it is often argued that mental representations are a needless cognitive addendum that organisms, by relying on their environment, can—and mostly do—avoid. Put differently, the idea here is that (a) extended cognition and representational cognition are opposed to each other—to the extent that cognition is extended, it should be thought to be non-representational, and to the

extent that it is representational, it should be thought to be non-extended—
and (b) most of cognition is extended and non-representational. However,
as I show in this chapter, these claims, at least as they stand, should not be
seen to be plausible. This is so for two reasons.

First, using the account of the evolution of cognitive and conative representational decision making laid out in the previous two chapters, I show
that there is no reason to think that internal mental representations are
in any way a needless cognitive addendum, or that reliance on them is
extremely rare. Second, on this basis, I show that there are in fact good reasons to think that a number of organisms will, at least sometimes, rely on
decision-making mechanisms that are both embedded in the environment
and representational. In fact, seeing cognition as representational may be a
prerequisite to fully understanding how and why it is sometimes extended.

The chapter is structured as follows. In the section below, I make more
precise how I understand the claim that cognition is (sometimes) extended.
Then, I show why it is mistaken to think that representational decision
making is a needless cognitive addendum whose evolution needs to be seen
as an exception to a general trend according to which most organismic
decision making is non-representationally driven. Next, I show why understanding the evolution of representational decision making can open up
new ways for understanding how and when it might also be extended. I
summarize my conclusions in the final section.

Extended Cognition

The key idea behind the recent work in extended cognition is that, while
cognition traditionally used to be thought of as an activity that is internal
to the organism (made possible virtually solely by its brain), it is in fact
much better seen as involving the organism's environment in key ways.
This idea has become known as the extended mind thesis (EMT in what
follows). There are two main ways of understanding this thesis (see, e.g.,
Rupert, 2009; Adams & Aizawa, 2008; Schulz, 2013c).

First, this thesis can be formulated as a metaphysical claim. According to
this version of the thesis, it is the case that mental states or processes are, at
least at times, partly or wholly realized by features of the organism's external
environment. For example, to the extent that the particular network of causes
and effects that (at least according to the major—functionalist—theories of

the nature of mental states) makes up a given mental state is located in the organism's environment, the relevant mental state itself is seen to extend into the environment (Clark & Chalmers, 1998; Sprevak, 2009). More generally: cognition, on this picture, just *is* the dynamical interplay between organismic action and environmental conditions—it spreads out beyond the head of the organism and reaches deep into its environment (Chemero, 2009; Stotz, 2010; R. Wilson, 2010; L. B. Smith & Thelen, 1994).

Second, the EMT can be formulated as a methodological claim. According to this version of the thesis, it is the case that, in *studying* cognition, it is necessary to situate an organism's cognitive processing within its wider environment (Adams & Aizawa, 2008; Rupert, 2009). More specifically, the idea here is that cognition cannot be fully or well understood unless it is placed in a wider framework that includes many parts of the organism's environment—such as the stimuli that kick-start many of the organism's cognitive processes, the external aides the organism uses to facilitate its cognitive processes, and the external circumstances that change as a result of the organism's cognitive processes (and which might then form the basis of further episodes of cognition). In short: the idea behind the methodological form of the EMT is that, independently of whether cognition actually is (partially) constituted by parts of the environment, it needs to be studied by taking into account the state of this environment (see also Schulz, 2013c).

Now, for present purposes, it is mostly the first, metaphysical form of the EMT that matters. That is, in what follows, the question is mostly one of whether all or parts of the cognitive processing that organisms undergo in fact happen outside of the skin of the organism. However, since it is plausible that the actual nature of various cognitive processes—that is, whether they are extended or not—has implications for how these cognitive processes are best studied, much of what follows likely will also apply, at least in part, to the (generally much less controversial) methodological form of the EMT.

Given this, it is further important to note that that the EMT—especially in the metaphysical version—is typically seen as providing some kind of antithesis to representationalist accounts of cognition (Brooks, 1991; van Gelder, 1995; Rowlands, 1999; Beer, 1990; L. B. Smith & Thelen, 1994). More specifically, it is often thought that commitment to the EMT leads to skepticism about the representational nature of cognition. Silberstein and Chemero (2012, p. 40) express this point as follows:

In extended cognitive science, non-linearly coupled animal–environment systems are taken to form just one unified system. This removes the pressure to treat one portion of the system as *representing* other portions of the system—at least for many cognitive acts. That is, if the animal–environment system is just one system, the animal portion of the system need not represent the environment portion of the system to maintain its connection with it. There is no separation between animal and environment that must be bridged by representations. So extended cognition invites anti-representationalism.

In the present framework, the key thought behind this anti-representational bias of the EMT can be spelled out as follows.

First, there is the fact that mental representations are ways of getting the organism's environment into its mind. As made clearer in chapter 2, they are higher-level mental states that contain information about the environment (at least to the extent that they are cognitive). So, representational decision makers can be seen to be organisms that react to the environment by first creating a representation of that environment in their mind, and then reacting to that representation.

Second, there is the fact that an organism can also use the environment directly to make a decision about what to do. This is easily seen by falling back onto the model of purely reflexive decision making laid out in chapter 2: organisms can just react to the state of the world as it is—by letting it trigger a behavioral response—and do not need to first create a mental representation of either the state of the world or of the best way of reacting to the state of the world. Their behavioral reactions might then change the state of the world, which will trigger further behavioral responses on the part of the organisms. In this way, decision making can be seen as a dynamical, non-representational system involving the organism and its environment, where each side effects changes in the other (see, e.g., Chemero, 2009; Brooks, 1991; van Gelder, 1995; L. B. Smith & Thelen, 1994).

Given these two facts, the supposition that cognition inherently involves the environment seems to lead quite naturally to the supposition that cognition is non-representational—for these two ways of making decisions seem to be diametrically opposed to each other.[2] To the extent that an organism relies on a mental representation to decide what to do, to that extent it is not relying directly on the environment (it is relying on a "stand-in" of the environment instead), and to the extent it is relying on the environment directly, it is not relying on a mental representation about this environment.

More than that, though: these two facts can also be used to underwrite the idea that *most* organismic decision making should be seen to be non-representational and extended. For if it is possible to use the environment directly in making decisions, then why would it need to be introduced into the mind via mental representations? Why would an organism build a mental model of the world and manipulate that, when it can manipulate the world directly—especially if representationalist decision making is relatively slow and takes much concentration and attention relative to non-representational decision making (Clark, 1997; Rowlands, 1999; Shapiro, 2011b)? In this way, it might be concluded that representational decision making should be seen to be a biological oddity, and not as the paradigm way in which organisms interact with their environment (which, as noted in chapter 2, is the approach taken in much of human and animal psychology and cognitive neuroscience; see, e.g., Clark, 1997, and Rowlands, 1999; see also Rupert, 2004; Thompson, 2007; Beer, 1990).

EMT Argument

This argument can be summarized like this (I shall call it the EMT argument in what follows):[3]

(a) representational decision making is costly (in terms of time, concentration, and attention);

(b) externalized, non-representational decision making avoids the costs of representational decision making;

(c) ceteris paribus, evolution by natural selection favors less costly over more costly decision-making processes.

Therefore:

(d) much decision making should be expected to have evolved to be non-representational and extended.

However, as I will make clearer in what follows, I think the EMT argument and its conclusion are not compelling. This is so for two reasons. First, as presented, the EMT argument is invalid, as there is a premise missing that notes the benefits of representational decision making. Second, though, I do not think that adding this premise should lead one to conclude that representational decision making is *never* extended—for there is a second premise missing that notes that it is not only possible, but in fact plausible, that much decision making is both representational *and* extended. Put differently, the implicit presumption of the EMT argument that representational decision making and extended decision making are opposites of each other is mistaken. The question of whether decision making is extended and the question of whether decision making is representational should not be seen to be two sides of the same coin, but as two separate questions. Consider these two points in more detail.

Mental Representations Are More Than a Needless Cognitive Addendum

As the account of chapters 5 and 6 makes clear, defenders of the EMT argument are right in thinking that there are numerous circumstances that do not favor the evolution of representational decision making. However, as this account also makes clear, this point must not be overstated: indeed, a thorough understanding of the evolutionary pressures shaping decision-making mechanisms shows that these pressures are often toward the reliance on mental representations. For this reason, it is false to think that representational decision making is a biological oddity that has rarely evolved. To see this more clearly, recall the main driving force behind the account of chapters 5 and 6.[4]

This driving force is the idea that the reliance on mental representations in decision making has evolved—where it has evolved—to make decision making more cognitively efficient overall. Specifically, mental representations enable an organism, first, to adjust more quickly and easily to a changed environment, and, second, to streamline their neural decision-making machinery by (cognitively or conatively) inferring the best response to the environment, rather than having to store this response. Of course, as also noted in chapters 5 and 6, these efficiency gains come with losses in terms of the time, concentration, and attention it takes to make a decision. However, the key point to recall here is that the arguments in chapters 5 and 6 make clear there is good reason to think the balance of benefits to costs is sometimes positive (i.e., it favors representational decision making). In turn, these two facts—that is, that cognitive efficiency is a key driver of the evolution of representational decision making, and that it has at least sometimes favored the evolution of representational decision making—matter for two reasons.

First, they make clear that mental representations are not a needless cognitive addendum whose evolution has to be explained away as an oddity. Rather, representational decision making (at least plausibly) has evolved and has been maintained for exactly the same reasons that non-representational decision making has evolved and has been maintained: namely because, in the relevant contexts, it enables an organism to make decisions in an overall cognitively efficient manner. Put differently: cognitive efficiency is a key driver of the evolution of *all* mechanisms of decision making—representational and non-representational alike.

Noting this is key, as it shows that seeing mental representations as a needless cognitive addendum is mistaken. Mental representations are not something that is only needed or wanted once the organism has all of its basic needs for efficient decision making satisfied, but rather something that enables the organism to fulfill its basic needs for efficient decision making. For this reason, they are also not generally avoidable (where this is understood to mean that they can be replaced without loss): efficient decision making can *depend on* mental representations.

The second reason why noting that cognitive efficiency plausibly is a key driver of the evolution of representational (as well as non-representational) decision making matters here is that it suggests considerations of cognitive efficiency may in fact *relatively often* be in favor of at least some forms of representational decision making. In particular, as noted in chapters 5 and 6, a number of organisms—namely those living in complex social, spatial, or causal environments—are likely to find it adaptive to make at least some decisions representationally. Because of this, it is not plausible to see representational decision making as a statistical outlier or a peculiarity of very few organisms (just humans, say).

Now, it is of course true that the account of chapters 5 and 6 also shows that there are many cases in which representational decision making is *not* adaptive. However, this point should not be overemphasized: there is no reason to think that non-representational decision making is somehow a biological norm that representational decision making violates. Put differently: while there is good reason to think that representational decision making is not a universal way of managing interactions with the environment, there is also good reason to think it is not a highly uncommon one. (Both of these points are also borne out by the empirical research presented in chapter 2.)

In this way, it becomes clear that the EMT argument is invalid. The reason for this is that there is a key fact that is being suppressed in this argument: namely, that representational decision making also has substantive benefits. If we add a premise to the argument to take this point into account, conclusion (d) no longer follows. So, if we add

(a′) representational decision making has benefits relative to non-representational decision making.

to (a)–(c), the EMT argument becomes invalid: for whether representational decision making will evolve then depends on whether its benefits are greater than its costs. Furthermore, as noted in chapters 5 and 6, it is in fact plausible that the balance of costs to benefits quite often favors the evolution of representational decision making.

However, this does not mean that, where representational decision making *has* evolved, it should always be expected not to depend on the environment at all. In fact, the entire alleged contrast between representational and extended decision making is a false one: there is a further premise missing from the EMT argument. The next section brings this out in more detail.

The Evolution of Extended Representational Cognition

The account in chapters 5 and 6 can do more than show that it is false to think decision making should nearly always be expected to be non-representational: it can also show that decision making should frequently be expected to be representational *and* extended. To bring this out, I start by laying out three different ways in which the adaptiveness of representational decision making can be increased by supplementing the latter with the right sorts of environmental resources. Given this, I then consider the circumstances in which these resources are available (i.e., I specify the sorts of environments in which extended representational decision making is likely to be adaptive).

Three Ways of Adaptively Externalizing Representational Decision Making

To see why there is reason to think there are sometimes adaptive pressures toward the evolution of extended and representational cognition, recall the main downside of representational decision making: the fact that relying on mental content in making decisions is computationally intensive, and thus that it can be slow and require much in the way of concentration and attention. For this reason, any way of lightening these cognitive demands is likely to be adaptive: it reduces the costs of representational decision making while leaving the benefits intact.[5]

This can best be seen by recalling (from chapter 2) that the costs that come from representational decision making plausibly are an *increasing function* of the complexity of the decision rule or cognitive representational inference the organism relies on. In particular, the more complex this rule or inference is, the costlier relying on representational decision making is likely to be—and thus, the less likely it is to be adaptive. As a result, organisms that find ways of simplifying the calculation of these rules or inferences—without compromising their accuracy or reliability—are likely to have an adaptive advantage.

Given this, the key point to note here is that one way of simplifying representational decision making is by drawing on the resources of the organism's environment. More specifically, there are three main ways in which organisms can use environmental resources to make representational decision making faster and less needy in terms of attention and concentration.[6]

First, the organism could externalize its representational decision making nearly completely: it could rely on environmental resources to do most of the inferential work underlying representational decision making, and then just tap into these resources whenever needed to obtain the outcome of this work. To understand this better, consider a non-conative representationally driven organism with a long table of behavioral dispositions, many of which instantiate patterns that could be exploited by using a conative representational decision rule.[7] Now, as made clearer in chapter 6, if this rule is very complex to use, the net adaptive value of conative representational decision making would be relatively low here (up to the point that the organism would be better off if it remained non-conatively representationally driven).

However, assume now that the organism could transfer its table of behavioral dispositions onto an easily accessible and usable external object—a sheet of paper, say. If so, then the organism could just rely on the decision rule "Do whatever it says on the appropriate part of this sheet of paper." This rule can provide all the benefits that conative representational decision making provides in general: the organism does not have to store "internally" a long list of behavioral dispositions, and instead can just "compute" the relevant behavioral response to the state of the world. However, importantly, this rule is also very easy to apply—for the relevant "computation" here just consists in looking up the right behavioral response on the relevant sheet of paper. Importantly, this rule is also just as accurate and reliable as the original non-conatively representationally driven behavior—after all, it effectively mimics it. In short: organisms that are able to transfer and easily use an externally stored table of behavioral dispositions are well advised to do so—for they gain the benefits of conative representational decision making at a fraction of its cost.

Note that this kind of case represents an interesting combination of non-conative and conative representational decision making. On the one hand, this is just like standard conative representational decision making—the organism *infers* what to do by consulting a behavioral rule, instead of deciding what to do by consulting an internally stored table of behavioral dispositions. On the other hand, though, much about the above case is very much like non-conative representational decision making: in particular, at its core, the representational inference the organism is making reduces to reading a stored table of behavioral dispositions; it is just that this table is now externally stored. For these reasons, this sort of scenario seems to provide the best of both worlds to the organism: the organism can ease its adjustment to new environments and streamline its neural decision-making machinery by relying on a representational inference about what to do, but the inference itself can be nearly as fast and effortless as a non-conative representational decision about what to do. (It is only "nearly" as fast and effortless as non-conative representational decision making, as reading an externally stored table is likely to be slower and more effortful than consulting an internally stored one—given the different kinds of reading processes involved here—and the reading itself is still mediated by a conative representational inference. I return to this point below.) As long as this solution is available, therefore, there is good reason to expect it to be adaptive.

The second way in which a representationally driven decision maker could adaptively "outsource" some of the necessary computations is by using a simplified form of the relevant decision rule, relying on the environment to do some of the necessary cognitive representational inferences of the original (non-simplified) decision rule. So, for example, assume it is adaptive for an organism to make its foraging decisions dependent upon whether the rate at which it recovers food from its current locale is lower than the rate at which it could recover food from the average locale in the area (see, e.g., Kacelnik, 2012). Now, it might be that there are two easily and constantly visible external signs (sign A and sign B) that note, respectively, the food-recovery rate of the current patch and the average food-recovery rate in the area.

If this is the case, then the organism, instead of relying on the complex decision rule:

(C) "compare the rate of recovering food at the current locale with that of recovering food from the average locale in the area, and stay if and only if the former is greater than the latter."

could just rely on the much simpler rule:

(S) "stay if and only if the number on sign A is greater than the number on sign B."

On the reasonable assumption that comparing the numbers on signs A and B is more easily done than estimating and comparing the relevant food-recovery rates "from the ground up," it will again be true that the organism can realize all of the benefits of cognitive and conative representational decision making without having to pay much in terms of lost decision-making time or increased need for costly cognitive resources.

This sort of scenario is noteworthy, not for combining non-conative representational and conative representational decision making (as in the previous case), but rather for combining cognitive and non-cognitive representational decision making. In particular, in this case, the organism makes decisions about what to do by using a representational inference that, in turn, is based on the outputs of two other, external inferences—which may or not be cognitively represented. In this way, instead of calculating something very complicated (the relationship of two estimated food-recovery rates) it calculates something quite simple (the ordinal relationship of two pre-calculated numbers). For this reason, this sort of way of making representational decisions should also be expected to evolve frequently—as long as the right kinds of environmental resources are available.

The last case worth mentioning here concerns a situation where a representationally driven organism does not externalize (parts of) the content of the computations it is engaging in, but merely uses the external environment as an aid in its *making* the relevant computations. To continue with the previous example, assume that, for one reason or another, the organism really does compute the complex decision rule (C), but, as it makes the relevant calculations, writes down intermediate steps on a piece of paper instead of holding them in its head. For example, it might estimate and write down the food-recovery rate of the current patch, and then do the same for the food-recovery rate for the average patch in the vicinity; it might then compare the last two rates and act accordingly. (Various more realistic examples of this kind could be conceived: for example, organisms might leave pheromonal or other markers that represent stages of the computational processes, which they can then later integrate to come up with a final decision—see, e.g., Rathbun & Redford, 1981.)

Now, the key point to note about this example is that externalizing decision making in this way can also significantly lighten a representational decision maker's cognitive load. By not having to store temporary elements of the relevant calculations, the organism's decision problem can be solved more quickly and with less recourse to concentration and attention (see also Kareev, 2012, and the discussion of instrumental conative representations in chapter 6). However, it is furthermore important to note that this way of lightening the organism's cognitive load is also the weakest of the three just mentioned—that is, it provides the smallest reduction in this cognitive load. The main reason for this is that the organism still has to do all of the complex calculations in question by itself. While this way of externalizing cognition thus streamlines the way the organism makes decisions, the fundamental nature of the organism's decision process remains unchanged.[8]

In short: there are three main ways in which an organism could adaptively externalize its representational decision-making processes: it could rely on externally stored tables of behavioral dispositions, it could rely on externally computed inferences, or it could rely on externally stored calculation aids for its own internal inferences. While these ways differ in the extent to which they provide adaptive benefits to the organism—with the first two providing significantly more benefits than the latter—they all provide some such benefits. Consider, therefore, the kinds of environments in which these different strategies for externalizing cognition are in fact realizable.

The Environments in Which Externalizing Representational Decision Making Is Adaptive

Given the fact that the above ways of externalizing and simplifying representational decision making are adaptive in environments that make the needed external resources available, it becomes important to ask when, exactly, these resources are in fact likely to be available. Put differently: which environments allow for the realization of these ways of adaptively externalizing and simplifying representational decision making? To obtain an answer to this question, it is best to begin with the third and least radical way of externalizing representational decision making, and then consider the two more radical ways after that.

Now, when it comes to the third sort of externalized representational decision making—that is, the reliance on features of the external environment as aids in the organism's representational inferential processing—there are no especially stringent conditions that need to obtain for this to be available. Many features of the external environment will have the minimal degree of permanence required to serve as temporary information stores for intermediate elements of an ongoing cognitive process. Of course, this does not mean that all features of the external environment can be used as temporary external information stores for all sorts of information (for example, storing the results of a calculation in a highly entropic system like a puddle of water is likely to be hard). The point is just that it is plausible that, for many representational decision makers, some features of the external environment are likely to have the needed properties to function as temporary information stores.

For this reason, it is plausible to think that many representationally driven organisms can and will rely on this kind of externalized decision making. This is useful to note, as this hypothesis also seems empirically plausible: humans, at least, seem to rely much on external aids when making many kinds of decisions (see, e.g., Rowlands, 2010; Griffiths & Stotz, 2000; Clark, 1997, 2008).

However, it is also useful to note that, since the adaptive benefits of this kind of externalization of representational decision making are likely to be relatively small—due the fact that the efficiency gains it brings are relatively small—it is unlikely that cases in which representational decision making is not adaptive (due to its high price in terms of lost decision-making speed, concentration, and attention) can be transformed into cases where representational decision making is adaptive. In other words, the evolution of representational decision making *in general* is unlikely to be impacted much by the possibility to externalize decision making in this way; this way of externalizing representational decision making can improve the efficiency of the latter, but probably not to an extent that this way of making decisions becomes adaptive where it is not already.

By contrast, when it comes to the two more radical ways of external-izing some of the demands of representational decision making—that is, the permanent external storage of the outcomes of some or even all of the needed representational inferences—there are some good reasons to think that the environmental resources needed to make this possible and adaptive are often unlikely to be available. This comes out quite clearly if it is noted that, if these externalized ways of making representational decisions are to be adaptive, the environment needs to provide resources satisfying the fol-lowing two conditions.

First, the external storage the organism proposes to rely on—that is, the feature of the external world that holds the relevant table of behavioral dispositions or the relevant outcomes of subsidiary inferences—needs to be *reliably accessible* (see also Clark & Chalmers, 1998). This is because, for the organism to consistently be able to make decisions about what to do, it needs to be able to consistently access the rule or inferences it relies on in its decision making. In turn, this means that, to the extent that these rules or inferences are externally stored, the external storage needs to be accessible to the organism whenever the organism needs it (which could be highly unpre-dictable). For organisms that are very mobile—such as many animals—this drastically restricts the kinds of external storage devices on which they can rely. For example, writing their table of behavioral dispositions on a rock is only useful if the organisms can see this rock—and, specifically, the writing on it—from wherever they are likely to find themselves. Also, the external storage device must not deteriorate too quickly across time. Writing a table of behavioral dispositions in the sand on a beach will, for many organisms, not be useful, as by the time the organism needs to use the table, it has likely washed away. (Much the same goes for externally stored parts of the relevant computational processes, as in the above second example.)

Second, the external storage device needs to be *efficiently usable*. This can be seen quite easily when considering an organism that consults an exter-nally stored table of behavioral dispositions. If this table is stored on a rock, say, and if this table is quite long, the organism needs to find a way to read this table quickly and efficiently—otherwise, many of the benefits of storing the table externally will dissipate. In particular, if the organism has no way to search the table for the entry that matches the state of the world (as it is represented or at least detected by the organism), and thus has to read the table line by line, it might take much too long to make a decision. In turn, this means that the available external storage needs to be organized in appropriate ways—that is, it either needs to be searchable, or organized in such a way that the organism can find the relevant information quickly without much searching (e.g., the information needs to be grouped in suitable ways).

Now, it seems clear that neither of these two conditions is easily satisfied. While some recent, human-designed devices—such as smartphones—would seem to fit the bill (many people carry their smartphones with them all the time, and these phones contain powerful hardware and software that makes it possible to store and find much information easily and quickly), few other features of the external environment are as reliably accessible and as easily useable as this. Because of this, for many organisms, there will not be many options with which to realize the potential of radically externalizing their representational decision making. However, importantly, this does not mean that there will be no such options. In particular, there is one important class of cases that, for a number of organisms, can satisfy the above constraints: namely, cases of "social externalization."

To see this, note that organisms that are social and highly cooperative can use the minds of other organisms to store information relevant to the first organisms' representational decision problems. This kind of externalized representational decision making can satisfy the above two conditions of reliable access and efficient usability, as organisms that are sufficiently social will be in close contact with each other, thus ensuring that they have reliable access to the information stored in each other's minds. Similarly, this information can be easily useable, as the external information store here is another *mind*—one of the most computationally efficient systems known (Churchland, 1985, pp. 458–461; Rumelhart et al., 1986; Clark, 1997). Also, there are often fast and efficient ways to communicate this information to other organisms (though I return to this point momentarily).

In this way, it becomes clear that externalization of representational decision making is a genuine option for some highly cooperative social organisms. However, there are two important caveats that need to be noted about the evolution of this kind of socially externalized representational decision making.

First, for socially externalized decision making to be possible, a quite extensive degree of cooperative sociality is required. In particular, organisms need to be in near-constant, fairly close proximity to each other, and they need to be motivated to provide information useful to one another (i.e., their biological interests must be sufficiently closely aligned to make extensive communication adaptive—see also Skyrms, 2010). Neither of these conditions is likely to be true for many organisms.

However, this does not mean that the degree of cooperative sociality required here is so high that *no* organisms should *ever* be expected to fulfill it. In fact, there is a small number of organisms for which it is plausible that they can satisfy this constraint. On the one hand, there are a number of organisms—including many primates, but also dogs, wolves, and some insects—that are in near-constant contact with other members of their group. On the other hand, there are reasons why, in some of these groups, extensive information sharing might be selected for: in particular, this will be so if this information sharing can be based on a division of cognitive labor (see also Skyrms, 2010).

To see this, note that if organisms cooperate by dividing their decision-making labor, their cooperation need not be evolutionary or psychologically altruistic.[9] In turn, this is due to the fact that, in cases of a division of labor, *all* of the organisms involved obtain a benefit. In the present context, this implies that, by dividing their decision-making labor, all of the organisms involved can make representational decisions more efficiently and quickly. For this reason, this kind of cooperation may be adaptive for many social organisms—which makes it more likely (though, of course, not guaranteed) that socially externalized representational decision making will be a genuine possibility that could evolve in these types of organisms.[10]

The second caveat to note concerning the availability of the resources needed for socially externalized representational decision making is that this kind of decision making depends on the fact that a relatively efficient means of communication exists with which the organisms can exchange the relevant information. In particular, organisms need to be able to communicate to other organisms the kind of information the latter require to make adaptive decisions—which may be highly specific and complex in nature (e.g., it might specify a particular sequence of behaviors to engage in, or a relatively exact number that forms the basis for a comparison with another fairly exact number)—and they might have to do so across some distance. For these reasons, the medium of communication the organisms rely on needs to be capable of the fast transmission of highly specific informational states. This constrains the possibility for socially external-ized decision making to evolve.

However, just as before, this does not mean that it makes the evolu-tion of this kind of externalized decision making impossible. In particular, natural (human) languages are one medium of information exchange that does have the needed features: it can express information of extremely high degrees of complexity and transmit it reasonably well over short to even medium distances (see also Pinker & Bloom, 1990). However, other options may exist as well. For example, organisms could *copy* each other—so that organism A follows the rule "do whatever B does in circumstances C_1" (and vice-versa for organism B and circumstances C_2)—perhaps with the model slightly exaggerating and slowing down the relative movements to make the copying easier (see also Sterelny, 2012). This, too, can transmit infor-mation of a reasonable degree of complexity across some distance. Another option is for organisms to rely on structured, non-linguistic forms of communication—such as bee dances or pantomimes (Russon & Andrews, 2010). Bee dances can represent the location and quantity of nectar in the area around a beehive with much accuracy (though they cannot express much else). If this is all the information needed by other bees to make deci-sions about where to forage next, then it becomes possible for bees to coop-erate by having some bees specialize in scouting for novel food sources, and other bees specialize in foraging at the food sources (see also S. R. Griffin et al., 2012; Seeley et al., 2006).

In short: the communication requirement of socially externalized repre-sentational decision making is steep, but not impossible to meet. Furthermore, among the organisms most likely to have the resources to meet it, humans stand out—human languages are some of the most sophisticated communica-tion devices in the biological world—but some other animals might qualify as well.

Putting all of this together, this thus suggests socially externalized rep-resentational decision making of even a radical kind will be a genuine evolutionary possibility—though only for a narrowly circumscribed set of organisms. Furthermore, the above suggests the organisms that are most obviously part of this set are humans: humans are highly social, cooperative, and have a highly efficient communication system at their disposal. As noted in chapters 5 and 6, they are also likely to have found representational deci-sion making to be adaptive. For these reasons, the above argument predicts that humans—and perhaps some other organisms—will frequently make representational decision in a socially externalized way. This is important for two reasons.

First, it provides a new source of support for the claim—put forward by defenders of the EMT (see, e.g., Clark, 1997, 2008; Rowlands, 2010, 1999; R. Wilson, 2010; Griffiths & Stotz, 2000; Stotz, 2010)—that extended decision making is widespread among humans. What the present argument adds to the considerations put forward by these authors is the idea that, while it is indeed plausible to think that much human decision making is extended, this does not mean that this kind of decision making needs to be thought to be non-representational. On the contrary: the representational character of this kind of cognition is key to its extended nature.[11]

In this way, the present argument both confirms and corrects the EMT. In particular, it adds to premises (a), (a'), (b), and (c) of the EMT argument the following further premise:

(a") The costs of representational decision making can, at times, be lessened if they are externalized.

From (a), (a'), (a"), (b), and (c), one can then derive the conclusion:

(d') Decision making should be expected to have evolved to sometimes be non-representational and extended, sometimes representational and extended, and sometimes representational and non-extended.

In this way, therefore, it can be seen that there is evolutionary biological support for externalized decision making after all.

Second, though, the above considerations also constrain the EMT somewhat, in that it suggests that, apart from humans, few other representational decision makers are likely to have found *radically* externalized representational decision making adaptive. Many other organisms either fail to be in sufficiently close contact to rely on a division of labor in their representational decision making or lack the necessary communicative abilities to use the results of this division of labor. In this way, the arguments here given also present somewhat of a counterpoint to much of the literature on extended cognition: while many organisms might rely on external resources as *aides* in their representational decision making (as the least radical way of externalizing representational decision making), fairly few organisms are likely to use external resources as major elements in the very *processes* underlying their representational decision making (as the two more radical ways of externalizing representational decision making).

Conclusions

I hope to have made a case for two conclusions in this chapter. First, I hope to have shown that a major argument for the idea that much decision making ought to be seen as extending into the environment and therefore non-representational is implausible.

Specifically, I have argued that it is wrong to suppose that the mere possibility of externalizing decision making entails that decision making should be expected to be non-representational. In fact, as the account of chapters 5 and 6 makes clear, there are a number of organisms that plausibly have found representational decision making to be adaptive. Relatedly, I have also argued that it is misleading to see representational decision making as a biological oddity whose evolution needs to be explained away: rather, both representational and non-representational decision making have been shaped by the same sorts of biological considerations, and neither one should be seen as a biological oddity.

Second, I have argued that extended decision making and representational decision making ought not to be seen as opposites of each other. Rather, many organisms are likely to rely on decision-making mechanisms that are both extended (in some form) and representational. In particular, many organisms plausibly find it adaptive to rely on external resources—including other organisms—at least as computational aids in their decision making, and some organisms might in fact rely on them as integral parts of their decision-making processes.

In this way, I hope to have advanced our understanding of both the representational and the extended nature of many decision-making mechanisms. Consider next the degree to which representational decision making—whether extended or not—ought to be expected to be specialist or generalist in character.

8 Specialist versus Generalist Decision Making

There is a set of theoretical and empirical disputes raging among students of both human and animal decision making. On the one hand, a number of scholars argue that human and animal decision making, at least to the extent that they are driven by representational mental states, should be seen to be the result of the application of a vast array of highly specialized decision rules: virtually every decision a representationally driven organism makes should be assumed to be based on a relatively simple decision procedure that is tailored toward that particular decision problem (see, e.g., Gigerenzer & Selten, 2001; Hammerstein & Stevens, 2012). By contrast, other scholars argue that we should see human and animal representational decision making as the result of the application of a handful general principles—such as expected utility maximization—to a number of specific instances (see, e.g., Stanovich, 2004; Kacelnik, 2012; Glimcher et al., 2005; Hausman, 2012; Chater, 2012). While much has been written concerning this dispute, the overall upshot is still quite unclear.

What I show in this chapter is that, using the results of chapters 5 and 6, it becomes possible to move this dispute forward. In particular, in what follows, I show that the account of the evolution of conative representational decision making defended in chapter 6 (together with the account of the evolution of cognitive representational decision making defended in chapter 5) makes clear that both sides of this dispute contain important insights, and that it is possible to put this entire dispute on a clearer and more precise foundation. Specifically, I show that differentially general decision rules are differentially adaptive in different circumstances: certain particular circumstances favor specialized decision making, and certain other circumstances favor more generalist decision making.

To bring this out, I begin by laying out the nature of the dispute. In the following section, I argue that, just considering insights from work in social and cognitive psychology, the dispute is still far from resolved. Next, I critically discuss Gigerenzer et al.'s evolutionary biological arguments for the specialized, simple heuristics-based view. Then I apply the insights of chapters 5 and 6 to the dispute concerning specialist and generalist decision making. Finally, I take stock and present my conclusions.

Two Views of Conative Representational Decision Making

While there are some important differences in discussions of the psychology of decision making in the human and non-human contexts, a major focus of both of these sets of discussions has come to be the question of whether representational decision making should be seen to be specialist or generalist in character (seemingly for independent reasons—though see also Kacelnik, 2012). Consider the two poles of these debates in more detail.

According to the generalist approach, conative representational decision making is typically based on decision rules that are informationally unrestricted (they take into account all of the considerations available to the agent), optimizing (they determine the best solution, given these considerations), and general (they work in the same way in all circumstances; for more on this, see, e.g., Gigerenzer & Selten, 2001; Fodor, 1983; Stanovich, 2004). Put differently, the generalist view has it that organisms make decisions by applying the same general, complex, and optimizing decision rule to the many particular circumstances they find themselves in.

In the human case, this approach is most clearly embodied in the views of a number of economists, psychologists, and neuroscientists: specifically, they argue that rational choice theory (RCT) gives an (idealized) account of how agents make decisions (see, e.g., Binmore, 1998; Hausman, 2012; Glimcher et al., 2005). (In the animal case, defenders of optimal foraging theory use a similar formal framework to argue that animals make decisions by maximizing some quantity that tracks the organism's inclusive fitness—see, e.g., A.I. Houston & McNamara, 1999; Trimmer et al., 2011; van Gils et al., 2003.) In particular, according to these economists, psychologists, and neuroscientists, organisms form a number of cognitive representations about which courses of action are open to them, about what the consequences of these different possible actions are in different states of the world, and about what the probability distribution is over the occurrence of the different states of the world. They then evaluate the different consequences of their possible actions: they assign them values that express how strongly they are committed to bringing about these consequences. Finally, they combine all of this information by averaging (in line with their probability assessments) these evaluations and choosing the highest such average.[1]

Note that this is meant to describe the psychology of decision making: while economists often also appeal to purely predictive or normative interpretations of RCT, what is at stake here are views that see RCT as actually describing the mental processes that are going on when people make decisions (see, e.g., Nichols & Stich, 2003; Glimcher et al., 2005; E. Fehr & Camerer, 2007; Colombo & Hartmann, 2017).[2] Furthermore, for present purposes, it is not the psychological accuracy of RCT per se that is important—everything that follows applies equally well to other generalist decision procedures (see, e.g., Kahneman & Tversky, 1979; Loomes & Sugden, 1982). What matters here is just that, on the generalist picture, decision making is thought to always operate in the same way—namely, by applying a general (and typically optimizing) decision rule to the particular decision problem the organism faces.[3]

By contrast, Gigerenzer et al. put forward a very different perspective on conative representational decision making (see, e.g., Gigerenzer & Selten, 2001).[4] In particular, Gigerenzer et al. claim that decision making should be seen to be modulated by "simple heuristics"—basic rules that are easy to apply and that make for quick, but often still quite accurate, decisions. Specifically, they claim that, instead of combing through all the information available to them, most organisms should be expected to only consider a small subset of that information. Further, instead of determining the optimal point at which the search for more information should be stopped, most organisms acquire information only until an easily determinable threshold has been reached. Finally, instead of determining the optimal way in which the "found" information is to be used to come to a decision—for example, by maximizing some internal variable that tracks fitness—most organisms determine which action to engage in on the basis of merely crude assessments of the considered data. Two further points are key to note about these simple heuristics.[5]

First, decision making that relies on simple heuristics is, in some ways, diametrically opposed to RCT-like decision making: it is informationally restricted, satisficing, and specialized. In particular, simple heuristics differ from generalist optimizing principles like RCT by not even aiming at determining the best response to the environmental situation; rather, they aim at producing behavior that happens to be good enough in the case at hand— where "good enough" means that the costs from potential mistakes are outweighed by the benefits that come from making decisions quickly and with less recourse to costly cognitive resources like concentration and attention.[6] Furthermore, the particular simple heuristic used will typically be different for different circumstances: decisions in one context will be made using a different rule from decisions in another context, as each decision rule needs to be specialized to particular circumstances to ensure it yields behavior that is "good enough" (sufficiently adaptive) in these circumstances.

Second, simple heuristics work by exploiting some of the structure that is built into the organism's environmental circumstances. Put differently, these decision rules take advantage of the fact that there is a lot of information specific to a particular circumstance that can be used to make acting in that circumstance adaptive. So, for example, when betting on whether team A or team B will win a given sports match, it can be reasonable to just bet on the team that one has heard of before. While this will not always be the most successful decision, it will be a profitable way of proceeding if well-known teams are more likely to be successful teams (Gigerenzer & Selten, 2001; Kruglanski & Gigerenzer, 2011). In this way, reliance on simple heuristics can yield behavioral outcomes that are just as adaptive in the relevant situation than what would be obtained by relying on optimizing, RCT-like mechanisms instead.[7]

Before considering the arguments that have been given for and against these two approaches toward the nature of representational decision making, it is important to note that the dispute between them in fact occurs within a space spanned by two logically independent dimensions. On the one hand, there is the question of whether the organism should be seen to be using only one highly abstract decision rule, or whether it should be seen to be using very many specialized decision rules. This dimension thus concerns the *number* of decision rules an organism uses to make decisions (and thus, the generality of its decision rules—the fewer of these there are, the more general they will need to be). On the other hand, there is the question of whether the decision rule or rules the organism uses are optimizing in character, or whether they are merely "satisficing" (in a broad sense of this term) in character. This dimension thus concerns the *nature* of the decision rule or rules the organism uses to make decisions. These two dimensions are logically independent of each other, since it is conceivable that an organism uses only one satisficing decision rule to make all of its decisions, or that an organism uses many decision rules, each of which is optimizing in content (albeit optimizing over different variables).

However, in practice, there are some reasons to expect that satisficing and specialization, on the one hand, and optimization and generalization, on the other hand, are correlated with each other. In the main, this is due to the point just made: it is in the nature of satisficing that it works well only in a narrowly circumscribed set of environments. Put differently, since (as just noted) the success of simple heuristics derives from drawing on information specific to the relevant environments, it is plausible that organisms that rely on simple heuristics will need to rely on many different such heuristics, tailored to the specific information contained in the different environments it is acting in. Of course, it is *possible* that an organism can get away with relying on only one simple heuristic to make many different kinds of decisions (especially if it is in a position to accept some losses in decision-making accuracy); however, in general, it is plausible that reliance on simple heuristics will require organisms to adopt many different such heuristics for the different circumstances in which it needs to make decisions. On the flipside, this therefore implies that the more generalist a decision maker is—that is, the fewer rules it relies on—the less tailored to specific circumstances these rules will be, and the more optimizing in content they are likely to be. Because of this, the clustering of many satisficing decision rules and few optimizing decision rules, while not logically required, should be seen to have some theoretical plausibility.

Psychological Perspectives on the Dispute

Which of the two views about (human or non-human) conative representational decision making—satisficing and specialist or optimizing and generalist—is more plausible? It turns out that purely by considering the status of the empirical literature in social psychology, cognitive psychology, and economics, answering this question is not straightforward.

On the one hand, there is little doubt that humans (and non-human animals) at least sometimes make decisions using simple heuristics

(Gigerenzer & Selten, 2001; see also Stanovich, 2004; Chater, 2012). In particular, there is now ample evidence that humans often make decisions based on what (a) they know the best, (b) they perceive to be more popular, and (c) the first option they happen to have considered in their decision situation. Importantly, it can be shown that, while making decisions in these ways *generally* leads to highly non-optimal choices, *in some contexts*, making decisions in this way can be surprisingly accurate. For example, if people's knowledge of the options under offer can be assumed to track the "choice-worthiness" of these options, then their relying on the "recognition heuristic"—the rule that says "pick the option that you recognize"—can lead to successful choices (Kruglanski & Gigerenzer, 2011; Gigerenzer & Selten, 2001). Importantly, also, making decisions in this way can be shown to be much faster and easier—in terms of the needed concentration and attention—than calculating the optimal response using a more complex decision rule (Kruglanski & Gigerenzer, 2011; Gigerenzer & Selten, 2001; Kacelnik, 2012; Hagen et al., 2012).[8]

However, on the other hand, there is also quite a lot of evidence that humans (and other animals) at times interact with their environment by relying on highly general optimizing rules (Stanovich, 2004; Stanovich & West, 2000; Glimcher et al., 2005; Kacelnik, 2012). In particular, much work in neuroeconomics has shown that humans (as well as various other primates) seem to make at least some decisions by calculating something like the expected utility of the different options involved in the decision, and then maximizing over the results of this calculation (Glimcher et al., 2005; Platt & Glimcher, 1999; E. Fehr & Camerer, 2007). Equally, there is considerable psychological and behavioral economic evidence suggesting that at least some humans rarely rely on simple heuristics and instead mostly make decisions using something very much like RCT (Stanovich, 2004; Stanovich & West, 2000; Glimcher et al., 2005; Kacelnik, 2012; see also Binmore, 1998; Chater, 2012).

Altogether, what this implies is that it must be acknowledged that there is some ambiguity as to how successful the two accounts—the generalist, optimizing one and the specific, simple heuristics-based one—are. Some data speak to the general account, some to the heuristics-based one, and some data are hard to interpret (see also Stanovich, 2004; Binmore, 1998). For this reason, a conservative conclusion to draw here is that we are lacking a good sense of which decisions are likely to be based on more general optimizing decision rules, and which on less general, satisficing decision rules. Because of this, it seems clear that it is desirable to obtain more evidence—perhaps from other sources—about human and non-human animal decision making to push the dispute forward.

Evolutionary Biological Perspectives on the Dispute

Interestingly, it is indeed the case that there is another source of evidence often appealed to in this debate: evolutionary biology. More specifically, Gigerenzer et al. present two evolutionary biological arguments for the claim that much decision making should be expected to be based on simple heuristics (see, e.g., Gigerenzer & Selten, 2001).[9]

On the one hand, Gigerenzer et al. note that it will generally be adaptive to opt for decision rules that allow for faster and more frugal decision making: organisms are always better off if their decision making takes less time and requires fewer cognitive resources (Gigerenzer & Selten, 2001; Sadrieh et al., 2001; Gigerenzer, 2008). Further, they note that simple heuristics are tailored to allow for fast and frugal decision making: they place strong restrictions on the amount of information they consider and on how thoroughly that information is processed; in turn, this cuts down the time and resources it takes to compute the right decision (see also Todd, 2001; Sadrieh et al., 2001). For this reason, Gigerenzer et al. argue that reliance on simple heuristics should be expected to be more adaptive than reliance on RCT-like decision-making mechanisms.

On the other hand, Gigerenzer et al. argue that evolution by natural selection should, in general, be expected to lead to specialization. Over time, we should expect organisms to consist of a large array of specialized solutions to particular problems (Todd, 2001; see also Tooby & Cosmides, 1992; Millikan, 2002). Since exactly this is true for simple heuristics—they are decision rules specialized to particular circumstances—this thus makes for another reason to expect the evolution of decision-making mechanisms based on rules of this kind.

However, as presented, neither of these arguments can be seen to provide unequivocal evidence in favor of specialized, simple heuristics-based decision making.[10] The account defended in the next section will make this clearer, but at this point, two points can be noted already.

First, whether simple heuristics really come out on top by allowing for increased decision-making speed and frugality is not obvious. On the one hand, this overlooks the fact that organisms could also make decisions in a non-conative representational way: they could just store, for every situation they might find themselves in, what they ought to do in that situation. As noted in chapters 2 and 6, this is likely to be even faster and more frugal than relying on simple heuristics. Given this, it is not obvious that the adaptiveness of fast and frugal decision making really favors simple heuristics, rather than non-conative representational decision making. On the other hand, as will also be made clearer below, there are also benefits that come from grouping different decision problems under one more general decision rule: this allows for easier adjustments to a changed environment and streamlines an organism's neural architecture. Given this, Gigerenzer et al.'s first evolutionary biological argument cannot be seen to be plausible: it fails to consider the possible benefits of both non-conative representational decision making and generalist conative representational decision making.

This last point also applies to the second of Gigerenzer et al.'s evolutionary biological arguments. Whether specialized decision making is more adaptive than non-specialized decision making depends on which decision problems it is adaptive to treat as distinct, and which as separate. So, if making decisions in one context (e.g., when it comes to food consumption) is optimally dependent on many of the same factors (e.g., the agent's degree of risk aversion and her discount rate for future benefits) as making decisions in another context (e.g., when it comes to the choice of coalitions to join) then it may be more adaptive to have decision making in both contexts be the result of a single rule. This is due to the fact that, in this context, both decisions *should* be solved in the same way, so that improvements in how one set of decisions is made is immediately transferred to the other set of decisions (see also Schulz, 2008).

In short: as presented, Gigerenzer et al.'s evolutionary biological arguments are not compelling. However, this does not mean that Gigerenzer et al. are wrong to think that considerations from evolutionary biology can be useful to push the debate forward. The next section makes this clearer.

The Evolution of Representational Decision Making and the Content and Number of Decision Rules

In what follows, I use the insights of chapter 6 (together with some of the insights of chapter 5) to suggest when we should expect conative representational decision rules to be satisficing and specialist, and when optimizing and generalist. Before doing this, though, it is important to make explicit something that was left implicit in the discussion of this chapter so far: namely, that the dispute between generalist-optimizing and specialist-satisficing decision making should not be seen as an either/or matter.

More specifically, this dispute should not be taken to concern the question of whether organisms rely on one generalist decision rule or many highly specialized decision rules. Rather, this dispute should be taken to concern the question of when organisms should be seen to rely on relatively more generalist decision rules and when on relatively less generalist decision rules. The reason for this "graded" and non-exclusive view of the dispute is that (as also noted earlier), it has considerable empirical plausibility: most organisms—humans included—seem to sometimes rely on relatively more general and sometimes on relatively less general decision rules (see, e.g., Stanovich, 2004). What is not clear is why this is the case, and when organisms should be expected to rely on which kind of decision rule—hence, answering *these* questions should be seen to be at the heart of the dispute here. (Moreover, as will also become clearer below, the account developed in this section can be used to further support this graded picture of conative representational decision making.)

In order to most easily lay out my account of when we should expect the evolution of relatively more specialist and when of relatively less specialist conative representational decision making, it is useful to begin by recalling a key insight of the account in chapter 6. This insight concerns the fact that the evolution of conative representational decision making is (very plausibly) influenced by the balance of two major adaptive pressures.

On the one hand, it matters how quick and frugal the organism needs to be in making the relevant decisions and how many different instances of the relevant decision situations it faces. This is due to the fact that there is little benefit in relying on conative representations when the relevant decision problems have few patterns in them that can be explicitly picked up and worked with by the organism—either because there are only a few behavioral options open to the organism (i.e., its table of behavioral dispositions is short), or because the relevant patterns are too computationally complex to work with given the time and resource constraints the organism faces. Put differently, conative representational decision making is not adaptive when it requires spending too many costly resources (time, attention, and concentration) in order to obtain its benefits (easier adjustments to changed environments and a smaller neural footprint of the decision-making systems).

On the other hand, it matters how many different decision situations the organism faces that are structurally related to each other—that is, how many of its decision situations are such that changes in the way it has to respond to one of these decision situations are correlated with changes in the way it has to respond to the others. This is due to the fact that, in cases where an organism faces many structurally related decision situations, it has an easier time adjusting its behavior to a changed environment, and gains cognitive and neural efficiency by only needing to store one behavioral rule to handle all of them, rather than many particular non-conative representational behavioral dispositions tailored to each specific decision situation. Put differently, the more it is the case that the decision situations an organism faces can be treated as instances of a more general problem, the more conative representational decision making can be expected to be adaptive.

Further, as also noted in chapter 6, it is possible to specify which kinds of environments favor conative representational decision making and which non-conative representational decision making. In particular, concrete examples of situations where non-conative representational decision making is likely to be adaptive include many basic motor decisions (see, e.g., Abernethy & Russell, 1987; Fischman & Schneider, 1985; A. M. Williams & Hodges, 2004). The reason for this is that these are cases where organisms have such strong needs to make decisions quickly and frugally that the benefits from conative representational decision making are unlikely to be great enough to make the reliance on this way of making decisions adaptive. By contrast, conative representational decision making is likely to be adaptive in certain complex social, spatial, and causal environments. These environments favor conative representational decision making, as, first, they allow decisions to often be made relatively slowly—organisms have time to plan and consider which coalitions to join, which partners to marry, which paths to a faraway point on the terrain to take, which kinds of prey to hunt and how, and which tools to build and how. Second, organisms in these environments are likely to (a) face many decision situations of a similar type, and (b) the individually best behavioral response to these different instances can, at least sometimes, be relatively easily computed from a given behavioral function (as in the example of the decision to join available hunting parties laid out in chapter 6).

Now, the key point to note about this account is that it is not just applicable to the question of when an organism should rely on a conative rather than non-conative representational decision-making mechanism to interact with its environment, but also to the question of *which kind* of conative representational decision-making mechanism—generalist or specialist—an organism should rely on in a given situation. In other words, the core elements of the account of chapter 6 do not just speak to the reasons for the evolution of conative representational decision-making mechanisms *tout court*, but also to the reasons for the evolution of different kinds—specialist or generalist—of conative representational decision-making mechanisms.[11]

To see this, begin by noting (as also made clearer in chapters 2 and 7) that the above costs and benefits of conative representational decision making are *positive functions* of, respectively, the need for fast and frugal decision making, and the ability to quickly adjust to changes in the environment and to streamline the organism's neural decision-making machinery. The greater the need for fast and frugal decision making, the less adaptive conative representational decision making is—and the greater the ability to quickly adjust to changes in the environment and to streamline the organism's neural decision-making machinery, the more adaptive conative representational decision making is. Pointing out this gradedness of the adaptive pressures influencing the evolution of conative representational decision making matters, as it suggests that there are sets of values of these two variables that are such that they agree on the fact that conative representational decision making is adaptive, but differ on what *kind* of conative representational decision-making mechanism is adaptive. In turn, this fact allows us to subdivide the environments in which conative representational decision making is adaptive into those that are likely to favor relatively more specialist decision rules and those that are likely to favor relatively more generalist decision rules. When doing this, the following points become clear.[12]

One of the best candidates for environments in which generalist conative representational decision making is likely to be adaptive are certain complex social environments: namely, environments that are constituted by (generally only distantly related) other organisms, and which require organisms to react to the behaviors, expectations, intentions, and roles of others so as to act adaptively. Note that the relevant environments here do not include all environments featuring other organisms. As made clearer below and in chapter 9, some interactions with other organisms—for example, helping offspring or fighting with conspecifics—are more adaptively managed by specialist decision rules. The point is that *some* social interactions—namely those involving medium- to long-term interactions with unrelated group members—are adaptively subsumed under a generalist decision rule. There are three reasons for this.

First, at least some social organisms—including humans and many other primates—have to make a myriad of social decisions in the course of their lifetime (and these decisions are adaptively extremely important). For example, the choice of mates, the choice of hunting and foraging partners, the choice of defensive partners, and the choice of caretakers and babysitters are key decisions many social organisms need to make over and over again. In particular, for many social organisms, much of their life consists in picking the right family units, hunting groups, babysitters, social roles, and so on, given the particular social circumstances obtaining at the time (who is where in the social hierarchy? who has which ties to which other organisms? etc.)—for it is these sorts of decisions that fundamentally determine their adaptive success (Sterelny, 2012).

Second—as also noted in chapter 6—many organisms can make social decisions *very* slowly, and with much recourse to concentration and attention. In particular, the kinds of social decisions just mentioned (picking mates, coalitions, family groups, etc.) are not decisions that have to be made and remade in the space of seconds, but rather something that can be considered in the space of days: because social environments often comprise the coordinated activity of many organisms, the timeframes in which social worlds change are typically fairly slow. Accordingly, social decisions can generally be taken quite slowly and with much use of concentration and attention.

Third, many (or even all) of these social decisions tend to be correlated, in that changes in the adaptive behavioral response to one social decision problem come with changes in the adaptive behavioral responses to the others. For example, it is plausible that for many organisms, ultimately, all social decisions revolve around an organism's place in its group's social hierarchy: this may influence who that individual can choose as a sexual partner, which hunting or foraging groups it can or must join, which professions are open to it, and so on (Sterelny, 2012, 2003; Whiten & Byrne,

1997; Humphrey, 1986). If this is so, though, then any change in that social hierarchy—for example, due to the arrival of a new group member from another group—will have implications for all of these decisions: a new, high-ranking group member changes who the focal individual can marry or befriend, what sorts of professions or roles it can occupy, and so on.

Given all of this, it is plausible that organisms that treat most or all particular social decision problems as instances of the same overarching social decision problem—"dealing with a complex social environment," say—will be more adaptive than those that rely on specialized rules for each of the different cases. This will be true for the same reasons that conative representational decision making can be fitter than non-conative representational decision making: it is overall more cognitively efficient.

In particular, by only having to store one decision rule, generalist conative representational decision makers are highly cognitively efficient: they have a particularly easy time dealing with changes in their social environment, and they can rely on a highly streamlined neural architecture. More specifically, since social decision makers have to solve many social decision problems, organisms that rely on specialized decision rules for each of these problems would need to store many such rules. Moreover, with the existence of many such rules comes the need for a mechanism with which to determine which decision rules are to be used in which circumstances: given that the different decision rules are tailored to different problems, the organisms needs to be able to select the "right tool for the job"—that is, pick the right decision rule to follow in the situation at hand.[13] This will also increase the complexity of the organism's decision-making machinery. All of this can be avoided by generalist conative representational decision makers: their decision making can be more streamlined. This is important, since (as noted in chapters 5 and 6) such increases in cognitive efficiency are likely to be biologically adaptive.

Furthermore, the losses that come from generalist conative representational decision making in social environments will often be low due to the fact that organisms do not need to worry much about slowed decision-making speed or increased need for concentration and attention. Since they can make most major social decisions relatively slowly and without major competing pressures for attention and concentration, the fact that generalist conative representational decisions might be harder to make—that is, require lengthier and more complex inferences—is not greatly problematic.

In short: it is plausible that social conative representational decision makers can adaptively see many of their particular social decisions as instances of a general decision problem that concerns "dealing with a complex social life." This is due to the fact that the upside of generalist representational decision making here is large and the costs are small: generalist conative representational decision making increases the ease with which an organism can adjust to changes in its social environment and streamlines its complex cognitive and neural decision-making system—and that without leading to major fitness losses due to the decreased speed and frugality of its decision-making mechanism.

By contrast, many of the other environments in which conative representational decision making is adaptive are unlikely to be as well suited to generalist decision rules. In particular, various problems of spatial navigation (such as traversing unfamiliar territories) and causal navigation (such as foraging for specific types of animals or selecting foraging or breeding territories) are likely to be better solved by relying on relatively more specialized conative representational decision rules. There are two reasons for this.

First, in these cases, the different decision situations are likely to be structurally uncorrelated with each other. This means that it is relatively unlikely that changes in the environment befall all of these different decision problems equally. So, for example, if (as in chapter 6) the organism has to navigate a complex spatial terrain to reach an important foraging spot, and the nature of that spatial environment changes (e.g., after an erosion that alters the physical structure of that terrain), then it is unlikely that the nature of the organism's foraging problem, once it reaches the relevant foraging site, changes as well. Similarly, if an organism has to solve two different foraging problems—one concerning whether to forage for a given foodstuff at patch A or at patch B, and another concerning whether to consume a given foodstuff now or save it for future use (see also Hagen et al., 2012; Kacelnik, 2012)— then changes in the way one of these problems should be solved is unlikely to affect how the other is to be solved. The reason for this is that these problems are constituted by factors largely unique to them, so that changes in these factors are relatively unlikely to "spill over" into other problems.[14]

Second, it is plausible that decision-making speed and frugality matter greatly for many spatial and causal decisions. While these sorts of decisions need not be taken as quickly and frugally as the kinds of motor decisions that are characteristic of non-conative representational decision making, they are still likely to need to be taken faster than many social decisions. Organisms cannot deliberate for days whether to consume or store a given fruit—for by then, the fruit may have spoiled. Moreover, they cannot afford not to pay attention to what is happening around them while making a food consumption decision, as other conspecifics may attempt to steal their food (Hare et al., 2000; Byrne, 2003).

Putting these points together suggests that many causal and spatial decisions are adaptively made by relying on specialist heuristics. Since specialist conative representational decision makers rely on the information already contained in a certain context, they can, in that context, make decisions significantly faster and more frugally than highly generalist conative representational decision makers can. This is important, since—as just noted—fast and frugal decision making is likely to be rewarded for many causal and spatial decisions. Further, since there is a lack of a correlation in many causal or spatial decision problems, it is not much harder to adjust to changes in the environment (compared to what is true for generalist conative representational decision makers), as these changes are likely to be isolated to specific decision problems. Finally, while it is true that specialist decision makers need to store more decision rules than generalist decision makers, it is plausible that this cost is often outweighed by the benefits that come from specialist conative representational decision making—after all, as just noted, the fact that more decision rules need to be stored does not affect the ease with which organisms can adjust to changed environments, and it does allow for faster and more frugal decision making.

In short: organisms are likely to treat many foraging, spatial reasoning, or other causal and spatial decision problems as distinct problems that require distinct, specialized solutions. This is due to the fact that, here, the upside of specialist conative representational decision making is likely to be great, and the downside small: relying on specialized decision rules allows the organisms to make decisions quickly and frugally—which is adaptive in these contexts—and comes with relatively few losses in terms of the ease with which the organisms can adjust to changes in the environment (though it is likely to increase the size of the organism's neural decision-making machinery somewhat).

All in all, this implies the following. We should not see all conative representational decision rules to be on a par: some of these rules are likely to have a wider "reach"—that is, to affect more and more varied decision situations—than others. In particular, social decisions are likely to be governed by a relatively more general decision rule, and many causal or spatial decision problems are likely to be governed by relatively more specialist decision rules. These differences stem from the same pressures that drive the evolution of conative representational decision making in general—just at a smaller scale (i.e., one within the set of cases in which conative representational decision making is adaptive). Five more points are important to note about this argument.

First, my account partially corroborates and partially departs from the arguments concerning these issues given by other researchers. In particular, my account agrees with the claim of many other researchers that many social decisions are unlikely to be an appropriate match for simple heuristics (see, e.g., Sterelny & Griffiths, 1999, pp. 331–332; Sterelny, 2003, pp. 208–209; Buller, 2005, p. 147; though see also Hurley, 2005). However, my account comes to this conclusion in a different way from how these other researchers get to it. Specifically, according to my account, the reason why social decision making favors reliance on generalist, optimizing decision rules is not so much that reliance on simple heuristics makes decision makers predictable (though this may be true), but that social decisions can often be taken slowly and with much use of concentration and attention, while rewarding increased ease of adjusting to changed social environments and increased neural efficiency. Put differently, while I agree with Sterelny, Griffiths, and Buller that it is hard to find simple heuristics that reliably yield adaptive behavioral outcomes in many social settings, this is not all that is problematic with reliance on simple heuristics in these settings: another major problem here is the fact that reliance on simple heuristics in those contexts is relatively inefficient.

Second, this account of the evolution of different ways of conative representational decision making shows that it is not right to see the evolution away from non-conative representational decision making as necessarily akin to an evolution toward highly general, optimizing, RCT-like decision making. While this may be the outcome in some contexts, it is also possible that the evolution of conative representational decision making just involves the reliance on a number of highly specialized, satisficing, simple heuristics. This matters, as it reinforces the idea that the evolution of conative representational decision making is not as difficult as it may at first appear. In particular, relying on conative representations in decision making does not need to imply that the organism jettisons all of the decision-making speed and frugality that non-conative representational decision making brings. Rather, there are ways of making decisions in a conative representational manner that are still relatively fast and frugal. This is important, as it further clarifies and supports the account laid out in chapter 6.

Third, my account here further underwrites the idea that the most reasonable position to take in the dispute between simple heuristics and generalist, optimizing decision making is one of compromise: instead of seeing all decisions as relying on *either* specialized simple heuristics *or* generalist, optimizing decision rules, it is evolutionary biologically plausible to see human and non-human animal decision making as pluralistically relying on both in different circumstances. In particular, conative representational decision rules of different degrees of generality are differentially adaptive in different circumstances.[15] Indeed, the present account can make clearer *which* situations are more likely to be approached in a generalist way, and which in a specialist way. In this way, the evolutionary biological perspective laid out here fruitfully combines with the existing social scientific work on this topic to suggest ways of obtaining a better understanding of the nature of (conative representational) decision making.

Fourth, the above arguments make clear that some of the classic examples that Gigerenzer et al. rely on to illustrate and support their simple heuristics-based view of decision making are in fact quite misleading. In particular, Gigerenzer et al. often suggest (see, e.g., Gigerenzer, 2007, pp. 10–13) that the way baseball players catch fly balls are good examples of the reliance on simple heuristics. In fact, though, this is not so—these cases are in fact highly atypical examples of the use of simple heuristics. The reason for this is that they represent complex, artificial motor skills that can be executed relatively slowly (fly balls are in the air for a relatively long time). Most other motor skills are likely to be automated, and thus are likely *not* to depend on simple heuristics—for example, baseball players are unlikely to rely on simple heuristics when catching line drives, and instead practice to have the appropriate behavioral responses be simply triggered by the perception of a line drive. Instead, better examples of decision making based on simple heuristics concern particular food consumption, foraging, or spatial reasoning problems.[16] Put differently: Gigerenzer et al. oversimplify the dialectical situation here—the adaptiveness of simple heuristics is not a monotonically increasing function of decision-making speed and frugality. Rather,

reliance on simple heuristics is adaptive in an intermediate zone where reasonably—but not overly—fast and frugal decision making is adaptive, and
where organisms face many instances of different types of decision problem
that are relatively uncorrelated with other decision problems.

A fifth and final implication of the account defended in this chapter
concerns the question of whether humans tend to choose "sub-optimally"
(given situationally appropriate standards of optimality). While the tradi-
tional answer to this question is affirmative (Tversky & Kahneman, 1974;
Kahneman, 2003; Kahneman & Tversky, 1979), some of the more recent
work on this question has come to a more ambiguous assessment (see, e.g.,
Jarvstad et al., 2014; Jarvstad et al., 2013; Stanovich & West, 2000). What the
arguments of this chapter make clearer is that the latter, intermediate position has some evolutionary biological support: there is reason to expect certain types of human choices to be optimal, but also reason to expect certain
other types of human choices not to be. While this does not amount to a
full resolution of the debate surrounding the optimality of human choices,
it is at least moving that debate forward.

Conclusions

I have tried to argue that the plausibility of generalized, optimizing decision
making—as compared to that of specialized, satisficing decision making as
well as that of non-conative representational decision making—can be pro-
ductively approached from an evolutionary biological point of view. More
specifically, I hope to have shown that, while the considerations typically
appealed to in this context—from social and cognitive psychology, economics, and evolutionary biology—do not settle these issues, the account
of chapters 5 and 6 can be used to make some progress here.

In more detail, I hope to have shown that specialist, satisficing decision
making is likely to be adaptive when an organism needs to make many different types of decisions representationally, but still relatively quickly and
frugally: reliance on simple heuristics enables an organism to reap some
of the benefits of conative representational decision making, without pay-
ing overly great costs in terms of lost decision-making speed and increased
needs for concentration and attention. Similarly, an organism is likely to
find decision making based on generalist, optimizing decision rules to be
adaptive if it can make decisions without major pressures toward speed,

concentration, or attention, and where these different decisions are structurally related, so that, by treating these different decisions as instances of the same overarching decision problem, the organism can save costs in adjusting to changed environments and decrease the size of its neural decision-making architecture. In short: we should expect conative representational decision making to be based on many specialized rules if there are patterns in the decision situations the organism faces that are better handled as distinct patterns, rather than as instances of one overarching pattern—and the reverse.

Interestingly, there is a further instance of the issues considered in this chapter that deserves to be considered on its own terms: namely, decisions about whether to help other organisms. The next chapter considers this in more detail.

9   The Psychological Underpinnings of Helping Behaviors

What motivates an organism to help another? This is still very much an open question, despite being quite widely discussed: several different answers have been proposed (both positive and negative), but none can be considered definite and generally accepted (see, e.g., Stich et al., 2010; Dovidio et al., 2006; E. Fehr & Gaechter, 2000; Batson, 1991; Nagel, 1970). Given this lack of a settled account of the psychological structures under-writing helping behavior, it is perhaps unsurprising that researchers have looked for new ways to investigate this issue. Among these new approaches is an evolutionary biological one: specifically, a number of authors have tried to assess the evolutionary pressures on different cognitive architectures with a view to their ability to lead to helping behavior (Sober & Wilson, 1998; see also Stich, 2007; Schulz, 2011c; Kitcher, 2011; Clavien & Chapuisat, 2013).

As I try to make clearer in this chapter, there is much that can be said in favor of this evolutionary biological take on the psychology of helping behav-ior.[1] However, as I also try to make clearer, making this evolutionary biological approach fully plausible requires shifting the focus of the analysis away from the reliability of different mind designs to lead organisms to help others—which is what existing analyses have tended to concentrate on—and toward the cognitive efficiency of different mind designs for helping others—that is, the kind of perspective that this book has been concentrating on.

To show this, I begin by making clear how I propose to understand some of the key terms of the debate in the section below. On this basis, I first discuss the existing, reliability-focused evolutionary biological analyses of the psychology of helping behavior (in the following section), and then present a cognitive-efficiency–based evolutionary argument against psy-chological egoism (in the next two sections). I present my conclusions in the last section.

The Psychology of Helping Behavior

In the literature, it is common to understand an organism to be a psychological altruist if and only if it holds ultimate desires for the well-being of other organisms, and a (pure) psychological egoist if and only if it holds ultimate desires for its own well-being only (see also Sober & Wilson, 1998; Stich et al., 2010). However, while this way of understanding these two terms is indeed generally quite useful, some adjustments to and clarifications of these definitions are necessary here.[2]

First, in line with the remarks of chapters 2 and 6, the references to "ultimate desires" and "instrumental desires" need to be clarified and corrected. On the one hand, for reasons made clearer in chapter 2, instead of *desires*, I think it is clearer to phrase the issues here in terms of *conative representations*. On the other hand, while—as made clearer in chapter 6—it is possible to make sense of the distinction between "ultimate" and "instrumental" conative representations, I will not place much weight on this distinction here. In turn, this implies that, in what follows, I understand psychological altruism and egoism as follows:

an organism is a *psychological altruist* if its behavior at least sometimes originates in conative representations for the well-being of other organisms; and

an organism is a *psychological egoist* if its behavior only originates in conative representations for its own well-being.

Second, I here leave it open exactly what "well-being" consists in (Stich, 2007). The only assumption I make concerning this is that well-being is at least correlated with fitness: increasing an organism's well-being will tend to increase that organism's fitness (G. Brown et al., 2012, pp. 235–236). This assumption is quite uncontroversial, though, and is shared by most of the rest of the literature on this topic (see, e.g., Sober & Wilson, 1998; Buller, 2005).

Third, given the way altruism and egoism are understood here, there will (sometimes) be a major difference between the altruist and the egoist in the way they make decisions about whether to help someone else (this will only sometimes be the case, for, as will also be made clearer below, an organism need not be an altruist about helping everyone else—altruists are, in general, motivational pluralists). The egoist will always derive the helping behavior from her conative representation to further her own well-being:

she subsumes all helping decisions under this conative representation. By contrast, for the altruist, the decision whether to help turns on a competition among *different* conative representations: since the altruist is bound to have more than one conative representation—it is implausible that an organism would make all of its decisions based on a conative representation to help others (see also Sober & Wilson, 1998)—the decision which of these conative representations to act on cannot be subsumed under a yet further conative representation. Instead, as noted in chapter 6 and as will also be made clearer below, different situations need to *non-representationally* engage different conative representations to be the determinant of her actions (see also Papineau, 2003, pp. 117–118): the altruist decides to help S since her conative representation to do so "overrides" her conative representation to help herself. Figures 9.1 and 9.2 illustrate this difference (note that these figures make room for the existence of instrumental conative representations, but do not require them).

The fourth and final point to note concerning psychological altruism and egoism as understood here is that they are not exhaustive of the space of possibilities. In particular, there are two main ways in which an organism might be neither an altruist nor an egoist. On the one hand, an organism might be (partly) driven by (i.e., act on) conative representations that are neither for its own well-being nor for that of some other organism. For example, an organism might be driven by conative representations to make works of art, or to discover new sources of food—which (at least plausibly) concern

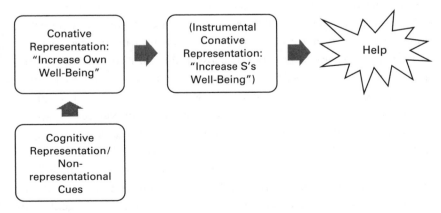

Figure 9.1
An egoistic decision to help organism S.

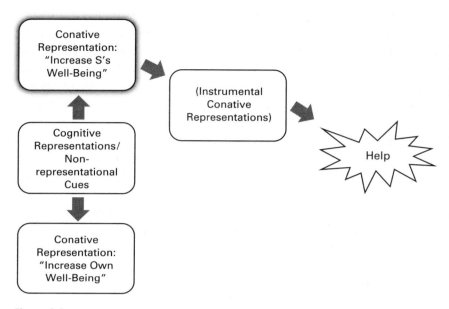

Figure 9.2
An altruistic decision to help organism S.

neither the organism's own well-being, nor that of other organisms, and so are neither altruistic nor egoistic. On the other hand, an organism might, in the relevant circumstances, be driven by non-conative representational behavioral dispositions only: as made clearer in the rest of this book, I take for granted that all organisms—whether altruistic or not—at least sometimes act only on the basis of behavioral dispositions that are not conative representationally derived.[3] If so, they are acting neither altruistically nor egoistically.

With the content of the theses of psychological altruism and egoism thus clarified, it next needs to be noted that it is still controversial which organisms should be seen to be (pure) egoists and which—if any—altruists. In particular, despite some ingenious attempts by various groups of researchers to settle these issues experimentally, the situation with regard to the spread and nature of psychological altruism is still not fully clear. While a full literature review of this work cannot be done here, the following remarks illustrate the kinds of difficulties encountered in this context.

A number of social psychologists have found that empathetic agents help subjects in need even if they have easy escape routes (such as no need to face the subject in need of help, the existence of prior commitments, etc.; Batson, 1991; Dovidio et al., 2006). While suggestive of the existence of altruism, this work leaves open the question of whether the helping occurs because empathetic agents genuinely care about the subject, or because they think (i.e., have cognitive representations with the content) that taking an easy escape route will not, in fact, alleviate their own personal distress at witnessing the subject in need (Cialdini et al., 1997; Stich et al., 2010).

Similarly, economists have found (among other things) that people engage in costly punishment of free riders (E. Fehr & Gaechter, 2000), that they make cooperative decisions as if they were bargaining with a partner (Misyak et al., 2014; Misyak & Chater, 2014), and that cooperation increases when the possibility for in-depth deliberation is restricted (Rand, 2016). However, this leaves open the motivations that lie behind these cooperative behaviors. Do people punish free riders or act in line with a form of "virtual bargaining" because they are committed to (a) fairness for its own sake; (b) helping others for their sakes (without, though, giving up on pursuing their own interests altogether); or (c) helping themselves (but also being aware of the fact that this may require compromising in some ways)? Similarly, that cooperation increases when deliberation is constrained could be either due to the fact that constraining deliberation shortcuts the agent's egoistic reasoning (and thus leads to less egoistic behavior), or that constraining deliberation prevents any kind of egoistic reasoning.

One of the most telling pieces of evidence in this context is the finding that both other-directed actions and self-interested actions engage the brain's reward centers (e.g., the ventral striatum), and that actions that require trading off benefits to the self with benefits to others are correlated with activation in parts of the brain associated with the monitoring and resolution among behavioral conflicts (the ventromedial prefrontal cortex and the anterior cingulate cortex; E. Fehr & Camerer, 2007). However, even here, ambiguities remain, since these findings do not make clear whether the conflict at stake is one among different behavioral goals, or different aspects of one behavioral goal (as would be true if people engaged in "virtual bargaining" of sorts, for example, or traded off short-term losses with long-term gains).

For these reasons, considering further evidence concerning these issues is useful. While the above work provides important data points to take into account in the explanation of the motivations underlying cooperative behavior—a fact to which I return below—the consultation of other findings is helpful to move this debate onward. In what follows, I suggest that an evolutionary biological perspective can provide such further evidence. Of course, in line with the discussion in chapter 3, the goal in this is not to end this discussion once and for all, but merely to add some further considerations that can improve our understanding of the motivations underlying helping behavior. Importantly, though, this modest, evidential role of evolutionary biology is still epistemically valuable, and it deserves to be taken seriously.

Reliability-Focused Accounts of the Evolution of Psychological Altruism

To date, the standard account of the evolution of psychological altruism is based on the reliability with which the latter leads to adaptive behavior. In relatively rough form, the idea here is as follows (for more detail, see Sober & Wilson, 1998; Stich, 2007; Schulz, 2011c).

Consider a case where it is adaptive for an organism O to help another organism S—for example, because S is O's offspring, or because doing so would induce S to reciprocate in the future, thus reducing O's risk of being left helpless. (Here and in what follows, I use "O helps S" as short for "O increases S's well-being.") Given this, for O to indeed decide to help S, O needs to (a) recognize that S is in need of help; (b) know what kind of help S needs; (c) be able to provide this kind of help; and (d) be motivated to provide this kind of help. So, how can (a)–(d) be accomplished? More generally, how can it be ensured that (a)–(d) are *reliably* the case?

Now, when it comes to (a)–(c), it is necessary to harmonize the needs of S and the perceptions, cognitive representations, and abilities of O. In general, this should not be taken for granted or assumed to be easily accomplished, but in at least some cases this kind of harmonization does seem possible to build up and maintain—indeed, the rest of this chapter can be seen to make part of the case for this conclusion. Because of this, it is sufficient for the present purposes to simply assume that (a)–(c) are satisfied, and focus on point (d)—that is, the question at stake here is just the one of how O can be reliably *motivated* to help S.

When it comes to (d), though, things are complex. In particular, assuming that O decides if and how to help S using conative representational mental states (an assumption to which I return below), two options are typically considered for how O might decide to help S: O could be an egoist that, through reasoning, comes to determine that increasing S's well-being is most conducive to increasing its own well-being (as in figure 9.1), or O could be an altruist and thus have a conative representation to increase the well-being of S that is triggered by the perception of S's need (as in figure 9.2).

It further has been argued that the altruistic solution is likely to be more reliable in leading O to help S—and thus (by assumption) more reliable in getting O to act adaptively (Sober & Wilson, 1998). Primarily, this is because there is less that can happen that would lead to O not being motivated to help S. In the egoistic case, there is a chance that O determines that doing something other than increasing S's well-being is most conducive to its (i.e., O's) well-being: perhaps the organism's cognitive representations change in such a way that it comes to think that taking drugs is most conducive to increasing its own well-being. However, if O does change its cognitive representations in this way, then it is bound to start acting maladaptively, since in this case, increasing S's well-being is (by assumption) adaptive. None of this, though, can happen if O is an altruist: for then, it will always be motivated to help S: after all, helping S is the content of one of O's conative representations.

However, it has been suggested that this argument underestimates the potential reliability of purely egoistic motivational architectures (Stich, 2007). This is due to the fact that, in order for a purely egoistic organism O to be a reliable helper of S, it is sufficient for O's help-inducing cognitive representation (viz., that the best way for O to increase its well-being is to increase S's well-being) to be "sub-doxastic" (Stich, 2007). Sub-doxastic states are cognitive representations that function much like other, standard cognitive representations, but are more rigid in that they are not amenable to updating and other forms of change—that is, they are fairly fixed components of an organism's mental life. Importantly, these states have also been appealed to in many other contexts, and their existence is quite widely accepted by now (Fodor, 1983; Carey & Spelke, 1996; Stich, 2007). Now, while there is more that can be said about these states, what matters most for present purposes is that they are immune to the worries about maladaptive cognitive representational changes mentioned above: after all, what makes these states unique is precisely the fact that they *cannot* be changed easily. Given this, it seems there is no reason to think that a sub-doxastically motivated egoist either could not exist or that it would not be just as reliable to help others as an altruist would be.

Now, it may be possible to respond to Stich's (2007) argument here (Schulz, 2011c). Fortunately, for present purposes, it is not necessary to assess the exact extent (if any) to which an altruist is more reliable than an egoist in helping another organism. The reason for this is that, as I try to make clearer in what follows, reliability is not all that is important for the evolution of altruism. There is also another evolutionary pressure— based on considerations of cognitive efficiency—that ought to be taken into account. Whether this other evolutionary pressure is the only driver of the evolution of the psychology of helping behavior or whether it acted in concert with the reliability-based adaptive pressures discussed in this section can be left open here (especially when keeping in mind the modest, evidential goals of the present inquiry).

The Evolution of Offspring-Focused Altruism

As I show in what follows, it is possible to apply the lessons of chapters 5, 6, and 8 to the debate surrounding the psychology of helping behavior. In particular, considerations of cognitive efficiency suggest that, when it comes to the generation of some kinds of helping behaviors, being altruistic can be more adaptive than being egoistic.[4]

To see this, it is best to begin by recalling the key conclusion of the previous chapter: it is adaptive for an organism to be a more generalist decision maker to the extent that the benefits of grouping several decision problems under one general conative representational decision rule (viz., the fact that the organism can adjust more easily to changes in its environment and streamline its neural decision-making machinery) outweigh the costs of doing so (viz., relative decreases in decision-making speed, and increased demands for concentration and attention). Recall further the flipside of this conclusion: namely, that being a more specialist decision maker is adaptive to the extent that treating a given decision problem as separate enables the organism to save sufficiently much decision making time, concentration, and attention to outweigh the greater difficulties of adjusting to changes in the environment and the increases in the size of the neural machinery underlying decision making that come with this. Recalling these conclusions is useful here, as they can be directly applied to the case of egoism and altruism.

In particular, there is a major kind of decision problem that, for some organisms, rewards being an altruist: helping offspring in need. To see this, begin by noting that the decision to help offspring in need has, at least for some organisms, two features that make it stand out from many other helping decisions (and indeed many other non-helping decisions as well).

First, many different kinds of organism—including many mammals—have offspring that are extremely vulnerable: they often do not have a fully developed immune system, their bodies are still quite weak, and their own energetic resources are greatly taxed by their own needs for growth. In such cases, delaying help that is needed can be greatly problematic: these offspring are unable to withstand or repair damage to their bodies, and even minor damage can become very serious if left untreated for significant periods of time (Smuts & Smuts, 1993; Cirulli et al., 2003; Rosenberg & Trevathan, 2002).

Second, the well-being—and thus, by assumption, the fitness—of an organism's current dependent children is often closely related to an organism's own well-being (and thus, by assumption, fitness). On a common definition, fitness is an increasing (probabilistic) function of the (expected) number of offspring—including grand-offspring—an organism has (Sober, 2001; see also Pence & Ramsey, 2013). Hence, increasing the well-being (fitness) of an organism's offspring generally increases the organism's own well-being (fitness) as well: the higher the well-being (fitness) of an organism O's offspring, the more offspring that offspring is likely to have—and thus, the more grand-offspring O is likely to have.

In fact, for many organisms, there is little else that competes with the importance of caring for their *current* offspring (at least when that offspring is young): specifically, it is a widely known fact about many mammals that the parent-offspring (and especially the mother-offspring) bond is exceptionally deep and of great importance for both the parent and the offspring (Chevrund & Wolf, 2009; Curley & Keverne, 2005; Churchland, 2011; Thometz et al., 2014). More particularly, for many organisms, conditions are such that helping their current offspring is often the primary adaptive demand on an organism—that is, it is an adaptive demand that cannot be traded off against anything else. For a concrete example, consider the case of sea otters (Thometz et al., 2014): these live in such hard conditions that offspring left to their own devices would almost certainly die. Moreover, these conditions are *reliably* harsh: it is not the case that sea otter parents could choose to let one season's offspring die in order to conserve resources that could be spent on offspring in future seasons when conditions are less harsh. This does not work, as future seasons are likely to be just as hard as this one, and sea otters cannot conserve energetic resources across seasons. In short: for many organisms, helping offspring is something that is worth doing at nearly every cost (Thometz et al., 2014).

Importantly, moreover, it is plausible that humans are among the organisms for which these two conditions obtain (Churchland, 2011).[5] In particular, human infants are extremely helpless and completely dependent on adult help (Rosenberg & Trevathan, 2002). Furthermore, there is generally little "seasonality" in the relevant environmental conditions: there is little reason to think that future offspring will be less dependent on help than current generations (largely due to the fact that infant humans are so completely dependent on adult help). Finally, often, "overproviding" help to human infant offspring is much less of a worry than not providing help when it is needed: the lost resources in the former case (food, time, energy, etc.) are generally small—individual helping episodes are typically highly constrained in space, time, and resources—but the cost of the former can be enormous—failure to attend to a child in need can make the difference between life and death of the child. In short: while being generally disposed to help their offspring is not adaptive for all organisms, it is so for some.

To avoid misunderstandings, two further points need to be noted. On the one hand, the claim is not that it is always adaptive for all organisms to help their offspring; indeed, the latter is known to be false (Trivers, 1974). The point is just that, for a significant number of organisms (including humans), helping their offspring is sufficiently often the adaptive response to make it adaptive to have a standing disposition to help their offspring.

On the other hand, even if it is adaptive for a certain population of organisms to be altruistically inclined when it comes to their offspring, members of the population need not always decide to help all of their offspring. This is due to the fact that, if an organism has multiple offspring in need of help, the decision how to apportion the available resources to these offspring is non-trivial. Indeed, it is consistent with an organism acting on a conative representation to help (all of) its offspring for it to at times decide to let some of its offspring die—for it may be that the best way to help its offspring is to give all of the available resources to some of its offspring, at the expense of some of its other offspring.

Moreover, something like this is arguably true in the human case, where infanticide is not uncommon (Hausfater & Hrdy, 1984; Hrdy, 1999). In particular, it is known that some human populations live in extremely harsh conditions in which keeping all offspring alive is often not possible; in such a case, a conative representation to help offspring may force human parents to make complex decisions about how to best allocate the available resources among the different offspring (Hausfater & Hrdy, 1984; Hrdy, 1999). Of course, this then sets up the potential for conflict among offspring and parents: all offspring have an interest in ensuring that *they* are the ones getting all the resources (see also Trivers, 1974); this further adds to the complexity of the parents' decision. However, the important point to note here is that all of this is consistent with the existence and adaptiveness of a disposition to always help one's offspring: it is just that acting on this disposition is often not straightforward.

Noting these points is crucial, as they suggest that, for many organisms, the decision whether to help offspring in need fits the circumstances in which specialized conative representational decision making is adaptive. First, helping offspring in need can reward fast and frugal decision making. Since, as noted above, the offspring of some organisms (such as many mammals) is highly vulnerable, the ability to decide to help these offspring quickly and without needing to spend much concentration and attention on this decision can greatly affect the well-being—and thus the fitness—of both these offspring and their parents (due to the importance of the fitness of the offspring for the fitness of the parents).

Second, for some organisms, the decision as to whether help offspring in need is responding to features that are different from those underlying other helping decisions—but is still highly situationally dependent. Since, as noted earlier, in many groups of organisms, the relationship between parents and offspring is unique (Churchland, 2011; Cirulli et al., 2003; Chevrund & Wolf, 2009), it is plausible that there is relatively little correlation among disruptions of adaptive responses to offspring helping decisions and disruptions of adaptive responses to other decision problems. Put differently, since the well-being of its offspring is so important to the fitness of an organism, it is reasonable to think that offspring helping decisions need to be sensitive to causal factors unique to them: even if the adaptive response to other (helping) decisions changes—for example, in cases of food scarcity or of increased dangers of predation while foraging—the adaptive response to offspring helping decisions will often remain the same (i.e., help the offspring). Of course, this is not to say that there is exactly one way to help offspring in need; in fact, there are normally very many different potential behavioral responses to an offspring in need (potentially including, as noted earlier, letting some of these offspring die).[6] What matters is just that these many different helping decisions form a class of their own.

Combining these two points implies—by the arguments of chapters 5, 6, and 8—that it is adaptive, for some organisms, to rely on a conative representation specifically focused on helping offspring in need. This is crucial, since the reliance on such a conative representation is precisely what psychological altruism amounts to. In particular, the key difference between the egoist and the altruist is just the fact that the former is a more general conative representational decision maker relative to the latter: the latter relies on specialized conative representations dedicated to helping others; the former treats helping decisions no differently from other decisions (e.g., foraging or mating ones) and makes them by relying on a very general conative representation to increase its own well-being.[7]

What we get here, therefore, is an efficiency-based argument for the evolution of psychological altruism in some groups of organism. The altruist is more adaptive, as it does not have to *determine* that helping the other organism is beneficial, but starts with this idea and can thus immediately assess how to best help that other organism. In this way, the decision to help is likely to be made more quickly with less recourse to concentration and attention in the case of the altruist—which, as just noted, is plausibly being rewarded (for some organisms) in the case of helping offspring in need.[8]

A final point that is worth noting about this argument is that I am here not presuming that either altruism or egoism is more *reliable* in leading the organism to help: despite the fact that the egoist is a more general representational decision maker than the altruist, it may still be no more likely to make the wrong decisions (e.g., because it is based on sub-doxastic states at the relevant junctions). My point is just that the mere fact that the egoist is a more general conative representational decision maker sometimes speaks against its adaptiveness relative to that of the altruist (independently of whether it is more likely to break down).[9]

An Extension: Helping Reciprocators

The argument of the previous section can sometimes be extended to other helping situations. A major example of this concerns some of the cases in which an organism has to decide whether to help reciprocators in need (see, e.g., Trivers, 1971; Sachs et al., 2004), where *reciprocators* are organisms that are disposed to repay help given to them (Skyrms, 1996, 2004). (Another possible example concerns organisms that live in group-structured populations in which helping group members in need is sometimes selected for— see, e.g., Sober & Wilson, 1998. However, since this introduces controversial issues that go beyond the main points of contention here—see, e.g., West et al. 2007, 2008—I will not discuss this example further, and concentrate on the case of reciprocation.) In particular, the decision to help a reciprocator has, in some circumstances, the same *sort of* features as the one to help offspring in need.

First, at least some decisions about whether to help a reciprocator need to be taken quite quickly: the help sought will be highly urgent. So, for example, what is at stake might be access to food resources that prevent starvation; waiting too long in making these decisions might render them moot, since the organism in need might then have starved (Carter & Wilkinson, 2013; Sachs et al., 2004). For the reciprocal relationship to get off the ground at all, therefore, significant amounts of decision-making speed are required. Of course, there is no reason to think that this will be the case for all reciprocal interactions—some of these can be decided on quite slowly—but it is plausible that this is so for some reciprocal helping decisions. This is likely to include some human ones, such as the decision whether to reciprocate help during battles or big game hunting. (I will return to this point below.)

Second, the decision to help a reciprocator generally responds to a unique set of factors and is thus likely to be uncorrelated with adaptive change in other helping decisions. In the first place, the degree to which organisms are related to each other, and the extent to which they can assess this degree of relatedness (Buss & Schmitt, 1993; Shackelford et al., 2005), are relatively unimportant to the adaptiveness of helping reciprocators, while they are crucial to many other helping decisions (Gardner et al., 2011; A. S. Griffin & West, 2002).

Further, the adaptiveness of reciprocal helping depends on the fact that the partners in the interaction interact multiple times, and that they are relatively unlikely to frequently interact with free riders (Skyrms, 1996, 2004). More than that, the relevant organisms need to be able to *keep track of* who helped whom in the past, and then relate that knowledge to current interactions, to be able to identify actual reciprocators. These features are also relatively unique to reciprocal helping (Skyrms, 1996, 2004; Hammerstein,

2003). Indeed, these features go some way toward explaining why, outside of the human realm, there are very few confirmed cases of reciprocation in nature, despite the fact that the latter would often seem to be adaptive (Hammerstein, 2003; Clutton-Brock, 2009; though see also Carter & Wilkinson, 2013).

In short: reciprocator-focused helping decisions plausibly form a reasonably closed domain with their own internal logic (see also Skyrms, 1996, 2004; Sachs et al., 2004). However, just as in the case of helping offspring in need, this does not imply that there is only one way to help a reciprocator (though, as will be made clearer below, this may be so in certain exceptional cases). In particular, reciprocal interactions can extend over long periods of time, during which the needs and abilities of the interacting organisms change drastically (Sterelny, 2012). For example, acquiring the skills needed to become a successful hunter or gatherer requires the help of other group members—but these other group members need access to unskilled labor, physical resources (like food), or social capital, such as reputation or status (Sterelny, 2012). For these reasons, some organisms face many different instances of the decision as to whether to reciprocate past help—but all of these instances are still aspects of the same *kind* of helping situation.

Given these features of (some of) the class of decisions to help reciprocators, the same argument appealed to in the previous section applies here, too. Specifically, the points just made, in combination with the arguments of chapters 5, 6, and 8, speak to the fact that it will be adaptive for some organisms to rely on a specialized conative representation for helping reciprocators.

First, organisms that have a conative representation concerning helping reciprocators can make the appropriate helping decisions more quickly and with less need for much concentration and attention than egoists. Egoists have to determine whether to reciprocate past help by deriving this from the conative representation to increase their own well-being. By contrast, organisms that already have a conative representation to reciprocate past help can make this decision much more quickly and easily. In turn, as noted earlier, this enables these organisms to make some of these helping decisions in a more adaptive manner.

Second, as in the case of helping offspring in need, having a conative representation for reciprocating past help does not drastically add to the organism's costs when adjusting to a changed environment. Since the decision whether to reciprocate past help often has its own internal logic, changes in the way the organism adaptively responds to other parts of its environment are likely to be independent of adaptive changes to the decision as to whether to help a reciprocator. However, since at least some organisms face many different instances of the decision whether to help a reciprocator, this is still a case that may be adaptively approached through conative representational decision making.

In short: by exactly the same reasoning as in the case of helping off-spring in need, an organism (partly) driven by a conative representation for helping reciprocators can be fitter than an egoist. As in the case of helping offspring in need, this is not due to the reliability with which the organism can engage in adaptive helping behavior; rather, it is a matter of the *efficiency* with which it can do so. Four further (related) remarks concerning this argument are worth making.

First, it is important to note that, as in the case of helping offspring in need, this argument does not apply to all organisms. However, there is good reason to think that the set of situations in which relying on a conative representation to help reciprocators is even more constrained than the set of situations in which it is adaptive to rely on a conative representation to help offspring in need. In particular, as noted above, while it is plausible that humans will find it adaptive to help needy offspring in a wide variety of situations, relying on a conative representation to help reciprocators would not be adaptive in cases where (a) organisms do not need to make the decision as to whether to help a reciprocator particularly quickly or frugally; or (b) they do not interact more than once with the relevant other organisms; or (c) they are in groups with large numbers of free riders; or (d) they are unable to keep track of whether past help was ever reciprocated.

However, this does not mean that there are no cases in which relying on a conative representation for helping reciprocators is adaptive. In particular, the above considerations are plausible for some *human* cooperative interactions: for, as noted earlier, at least some human cooperative interactions have the features that make decision making based on conative representations for helping reciprocators adaptive (E. Fehr & Gaechter, 2000; Falk et al., 2003; Sterelny, 2012). More than that: the above argument predicts that humans will have a stronger disposition to cooperate with others to the extent that this is something they have to decide on quickly and with little recourse to concentration and attention: for it is precisely in these sorts of circumstances that the conative representation to help reciprocators is most likely to be triggered. Importantly, this point is confirmed by the findings of Rand (2016) sketched earlier: cooperative tendencies increase the more cognitive load subjects are under when making the relevant decisions. For these reasons, the argument of this section, while not as widely instantiated as that of the previous one, should still be considered important—especially in the human case.

The second point to note concerning the above argument concerns the classification of an organism that acts on a conative representation to help reciprocators. On the face of it, this may seem to be an altruistic cognitive architecture: it is directed at helping a certain set of other organisms. However, against that, one might say that, really, the conative representation in question here is directed not at a certain group of other organisms, but at a certain form of behavior: reciprocation. While this behavior may be directed at others, and while it might result in increasing their well-being, the latter is not what the conative representation is about. If so, then this would not be psychological altruism, but something that is neither altruistic nor egoistic. Fortunately, settling this issue is not greatly relevant here: what matters is that this is a situation that favors the evolution away from egoism—whether it is toward altruism or a neutral form of a helping architecture is largely a matter of convention that can be left open here.[10]

Third (and very much in line with the discussion in chapter 8), this argument concerning the evolution of reciprocation-focused conative representations can, in some cases, lead to a different conclusion: namely, that the organism makes decisions most adaptively by relying on non-conative representational behavioral dispositions only (see also Jensen, 2012, p. 316). In turn, this would then represent an evolution away from egoism, but not toward altruism, for, as noted earlier, egoism and altruism presuppose the presence of conative representations, which do not underlie the organism's motivation to help here.

To see this, assume that the organism in question reciprocates help in exactly the same way at all times, and that this is also the only kind of help that is looked for.[11] So, for example, in the much-discussed case of food-sharing vampire bats (see, e.g., Carter & Wilkinson, 2013), it appears that the only way in which help is demanded or reciprocated is by regurgitating blood.[12] If that is so, though, then it is plausible that these bats do not need to rely on conative representations to make helping decisions: they can just react to the perception of their conspecific's need by *doing* the relevant behavior—that is, by regurgitating blood. After all, by the arguments given in chapter 6, conative representational decision making will only be adaptive if the organism needs to adjust its behavior to the state of the world—and do so in a principled way. Here, however, the organism always needs to respond in exactly the same way to all demands of help by reciprocators (e.g., by regurgitating blood)—there is no need to modulate this behavior at all. Given the account in chapter 6, therefore, it is reasonable to think that the organism is most adaptively structured if it avoids using conative representations to decide whether and how to help others, and just relies on non-conative representational triggers to engage in helping behavior. In short: it is also plausible that some helping decisions might be most adaptively taken by relying on non-conative representational behavioral dispositions only.

The fourth and final point worth making here concerns what happens when the argument of this section is combined with that of the previous section. In that case, the organism will rely on a complex, pluralistic, helping-oriented psychology: in particular, the organism would rely on specialized conative representations focused on helping offspring *as well as* on specialized conative representations focused on helping reciprocators (and on various further conative representations apart from these, too).[13] Given this, though, it is possible that the organism will, at times, face a clash between these different conative representations.

So, for example, assume it turns out that a pluralistically motivated organism is in a situation where its offspring *and* a known reciprocator are in need of help. By the above argument, the former fact triggers the conative representation to help the organism's offspring, and the latter fact triggers the conative representation to reciprocate (or it might even directly trigger a behavioral disposition to reciprocate in a particular way). Assuming—as will often be plausible—that providing both kinds of help at the same time is impossible, the organism thus faces a dilemma: which conative representation (or non-representational behavioral disposition) should be given priority in governing the organism's behavior?

However, as made clearer in chapter 6, while introducing some complexities, this issue does not change the fundamentals of the above argument. This is because the existence of several competing conative representations or non-representationally triggered behavioral dispositions only implies that there needs to be a way to resolve these kinds of conflicts. As noted in chapter 6, one way in which this could be done is by means of a "priority ranking" over the different conative representations and non-representationally triggered behavioral dispositions—for example, the altruistic conative representation to help the offspring could always trump the conative representation to reciprocate past help. Alternatively, this kind of conflict could be resolved with a priority ranking over the *outcomes* of all of the relevant behavioral dispositions: the organism might be disposed to reciprocate past help rather than help the offspring if the former has consequences of type A (e.g., much benefit at little cost) and little consequences of type B (few benefits at high cost). While this thus introduces some further complications to the present argument, the argument itself remains untouched. What is interesting to note, though, is that the findings reported in E. Fehr and Camerer (2007) are very much consistent with this idea: in particular, the fact that cooperative and non-cooperative behavior leads to activation in the areas of the brain associated with the monitoring and resolution of conflicts among behavioral goals confirms the prediction of the present account that humans, at least, are likely to rely on a number of specialized conative representations focused on helping others. This can thus be seen as an instance of "consilience": different pieces of evidence gain in epistemic importance when they can support each other.

Conclusions

Summing up, all of this leads to the following overarching conclusions. When it comes to evolutionary biological accounts of the psychology of helping behavior, it is useful to consider the situation from the point of view of what is most cognitively efficient, and not just of what is most reliable. When doing this, it becomes clear that there is adaptive pressure on at least some organisms to move away from being purely egoistically motivated: some organisms face pressure to act on conative representations to increase the well-being of their offspring, some organisms to help reciprocators in need, and some to do both. Two further points concerning these conclusions are worth bringing out explicitly here.

First, it is important to keep in mind that the conclusion here is that, for some organisms, it is maladaptive to make all helping decisions from egoistic motivations. This is consistent with it being the case that *some* of the helping decisions of virtually all organisms are adaptively taken with an egoistic mentality. For example, it may well be maladaptive to have a cona-tive representation toward helping strangers—for whether helping strangers is adaptive might be best determined by reasoning from a conative represen-tation to increase one's own well-being. Equally, the argument of this chapter is consistent with there being organisms that are pure egoists: for example, organisms—such as some reptiles—whose offspring are self-sufficient from an early age, and who face strong seasonality in the abundance of their food resources, might not find altruistic conative representations toward their offspring adaptive (in fact, they may find offspring cannibalism to be adap-tive: see, e.g., Polis & Myers, 1985). If they also do not live in groups of recip-rocators, these organisms might be most adaptively structured if they make all of their helping decisions by deriving them from a conative representation to increase their own well-being. The point of this chapter is only that this situation does not characterize *all* organisms *all* of the time.[14]

Second, the arguments of this chapter can be seen to further support the arguments from the previous chapter. In particular, the conclusions of this chapter underwrite the idea that it is plausible that conative repre-sentational decision makers do not make all decisions by deriving them from what maximizes their inclusive fitness (or the like), but that they will often rely on more specialized heuristics that focus on particular situations. In this way, some further evolutionary biological support can be obtained against the contention that all organisms make all decisions by deriving them from some one basic conative representation (as has been suggested by Schroeder, 2004; for critical discussion, see also Katz, 2005). In particular, together, this chapter and the previous one show that there are some rea-sons to think that this will not be true of all organisms—humans included.

Conclusions

It is now possible to draw together the different aspects of the book and thereby bring out several overarching conclusions. I have considered the evolution of a widespread suite of psychological traits: namely, decision making based on cognitive representations (multi-tracking representations about the state of the world) and conative representations (explicitly encoded instructions about what the organism is to do). I have argued that a major factor influencing the evolution of this representational way of making decisions is the fact that it can make decision making more cognitively efficient: in some contexts, representational decision making allows organisms to have an easier time adjusting to a changed environment and to streamline their neural decision-making systems.

While these efficiency savings are tempered by some losses in cognitive efficiency—representational decision making is generally slower and takes more in the way of concentration and attention—I have argued that this balance is sometimes positive. Specifically, organisms that live in certain spatial, causal, or social environments—namely, environments that require organisms to modulate their behavior to relatively complex combinations of different states of the environment, where these modulations can be derived from a number of relatively simple principles, and where fast and frugal decision making is not greatly important—are likely to find some form of representational decision making to be adaptive. Three further points are important to note about these conclusions.

First, they partially contrast and partially overlap with some of the existing work on the evolution of representational decision making. They overlap with this work in the sense that my conclusions and the others in the literature agree on the idea that complex spatial, causal, and social environments are key to the evolution of representational decision making. My conclusions contrast with the others in the literature in the sense that they differ over the reasons for why these environments matter: it is not that, in these environments, representational decision makers can do things that non-representational decision makers cannot; it is that they can do the same things better.

Second, my account of the evolution of representational decision making spells out a moderate, evidential view of evolutionary psychology. This view contrasts with versions of evolutionary psychology that are more radical, in that, instead of seeing the appeal to evolutionary biology as allowing for a drastically different (and deeper) understanding of human and non-human minds, it sees evolutionary biological considerations as just presenting evidence for certain psychological, anthropological, social scientific, and philosophical hypotheses—nothing more, but also nothing less. Put differently, what my account of the evolution of representational decision making shows is that such a moderate, evidential form of evolutionary psychology is both feasible and interesting.

Third, my account of the evolution of representational decision making can be used to make some progress in three debates concerning representational decision making. In particular, knowing something about the circumstances in which representational decision making is adaptive also suggests when representational decision making should extend into the environment and in which way, when it should be based on more specialized and when on more generalist decision principles, and when it should lead to altruistic helping behavior.

In this way, I hope to have shown that representational decision making is a fascinating trait whose evolution deserves to be more closely studied: despite being a rather abstract psychological ability, much can be said about why organisms have (sometimes) adopted it and how it works when they have done so. More generally, I also hope to have brought out the benefits of interdisciplinary studies of the mind: by combining evolutionary biology, psychology, anthropology, social science, and philosophy, we can acquire fascinating insights into the way human and non-human animals interact with the world around them.

Notes

Chapter 1

1. I understand "decision making" broadly: it is about how an organism determines how to behave. For this reason, there can be both representational and non-representational ways of making decisions. Those who prefer restricting the term "decision making" to representationally driven organisms should replace all occurrences of "decision making" in the context of non-representationally driven organisms with something else (such as "behavior determination"). Fortunately, nothing of substance hangs on this terminological choice here (see also Ramsey, forthcoming).

2. Note that the point here is not that all of these evolutionary biological hypotheses are necessarily true for all traits; the point is just that the fact that a given trait has all of these features explains what makes investigating the evolution of this trait particularly interesting. Note also that one might, of course, be interested in the evolution of a trait even if it is not complex, ecologically important, or widespread.

3. Some scholars have coined the term "philosophy of nature" to describe this sort of project (Godfrey-Smith, 2009; Sterelny, 2012). Others (see, e.g., Carruthers, 2006) prefer to go back to Locke, and describe this kind of project as being an "under laborer" of science. Recall also Sellars (1963, p. 1): "The aim of philosophy, abstractly formulated, is to understand how things in the broadest possible sense of the term hang together in the broadest possible sense of the term."

Chapter 2

1. I here focus on organisms, but nothing in the book precludes the possibility that there are also other kinds of decision makers—such as groups of organisms or even parts of organisms (such as groups of neurons—see, e.g., Fidgor, 2011). However, the latter set of claims requires a separate investigation of its own (see also Schulz, 2016b).

2. Relatedly, I also leave open what the nature is of the relevant intermediate state—for example, whether it is imagistic, an item in the language of thought, or something else.

3. There are many different versions of these accounts (see, e.g., Garson, 2014, chap. 7; Stegmann, 2009), but for present purposes, the above rough statement is sufficient.

4. This is also related to the point mentioned earlier that the distinction between representational and non-representational ways of making decisions may not be categorical, but merely graded.

5. Relatedly, I here do not *require* that mental representations satisfy Evans's "generality constraint" (G. Evans, 1982; for discussion, see, e.g., Camp, 2004; Carruthers, 2006) or that they are "systematic" (Fodor & Pylyshyn, 1988; Calvo & Symons, 2014; Aizawa, 2003; McLaughlin, 2009). However, as made clearer in chapters 5 and 6, it is plausible that, *as a matter of fact*, representational decision making is at least somewhat systematically structured.

6. As will be made clearer below, representational decision makers can also make some decisions non-representationally. The point is just that, to the extent that they make representational decisions, these decisions will be based on intermediate, higher-level mental states.

7. It is even debatable whether some plants can be put into this category (e.g., when it comes to the opening up of flowers when in sunlight), but I shall not consider this further here. See also Calvo et al. (2014).

8. Note that some (see, e.g., Dretske, 1988) might see the magnetosome as representational. However, as also noted earlier, this is not enough to make magnetotactic bacteria representational decision makers *in the sense that it is relevant here*.

9. Note that, as will also be made clearer throughout this chapter, this is a point about the kinds of explanations taken seriously in science: it is not a point about how behavior can be interpreted in principle. It may be true that all behavior can, in principle, be interpreted as being either representational or non-representational (see, e.g., Dennett, 1987, for more on this); the point here, though, is that, as a matter of fact, successful theorizing in some contexts is focused on one or the other.

10. Of course, many such behavioral dispositions may also be innately acquired.

11. Note that some of these regions are also involved in the representational decision-making system. I return to this below.

12. A different way of putting this point is that cognitive representations, as they are understood here, "chunk" the information the organism receives about their environment: they bind together different perceptual states. This mode of phrasing the issues will become important again in chapter 5.

13. Relatedly, the issues here also do not concern methodological principles like "Morgan's Canon" (Wilder, 1996; Sober, 1998b).

14. This is of course not to say that this consensus is universally accepted—for some criticisms, see, e.g., Chemero (2009); Hutto and Myin (2012); Gallagher (2015);

Brooks (1991); van Gelder (1995); Rowlands (1999); Ramsey (2007). However, as I try to show in what follows, this consensus has much that speaks in favor it. At any rate, I explicitly discuss some of these alternative views in chapter 7. Note also that the goal in what follows is not to say that all parts of cognitive science should be expected to be thoroughly representational: as will also be made clearer in chapters 5 and 6, there is in fact good reason to expect some parts of cognitive science—such as those dealing with fast motor decisions—to continue to avoid appealing to mental representations. The point is just that it is now widely accepted that, for *some* organisms, *some* aspects of decision making are representational.

15. This division is somewhat arbitrary, especially due to the fact that neuroscientific considerations are now standardly part of the work in the other sciences of the mind (a point to which I return below). Still, I think that this three-part division is useful in coming to see some of the major points being made in the relevant literature.

16. None of the work below claims that all organisms ought always to be seen to be representational decision makers. Indeed, all of the accounts sketched below are consistent with—or might even explicitly posit—the existence of some organisms that are purely non-representational decision makers. Equally, all of the accounts below can—and some explicitly do—make room for some decisions of representational decision makers to be non-representationally driven. The claim is just that some decisions of some organisms are now widely accepted to be representationally driven.

17. See also Tulving (1985) for an early form of this kind of theory.

18. As noted earlier, these two systems are often assumed to lead to conflicts as well: see, for example, Greene (2008) and Haidt (2001).

19. Gigerenzer and his colleagues also attack accounts of decision making that focus on something like expected utility maximization (see, e.g., Gigerenzer & Selten, 2001). I consider this aspect of their account in chapter 8 below.

20. An exception to this is the work of Linda Smith and Esther Thelen—see, for example, L. B. Smith and Thelen (1994). However, the extent to which this work can explain—rather than predict—infant development is highly controversial. See also Clark (1997, chap. 5).

21. I here do not focus on the many studies concerning the recognitional abilities of different animals, as it is often unclear whether these involve genuine mental representations or just relatively coarsely individuated physical (i.e., perceptual) states; see also Allen and Hauser (1991).

22. Also, van Gils et al. (2003) argue that red knots—a type of shorebird—are well seen to make foraging decisions by combining information about their current foraging patch with prior knowledge about the quality of other patches in a satisficing decision rule. (While their research does not explicitly use the term "representational," it

seems clear that, substantively, their model very much fits the account laid out in this chapter.) Similar results are in Langen (1999) and A.I. Houston and McNamara (1999).

23. These need not be seen as the only cognitive resources. However, for present purposes, it is sufficient to concentrate on these two.

24. For what follows in the rest of this book, it is sufficient if it is acknowledged that representational decision making, *to the extent that it is inferential*, is more concentration and attention-hungry.

Chapter 3

1. Some critics are also concerned that work in evolutionary psychology can be used to underwrite dubious moral or political viewpoints (see, e.g., C. Fehr, 2012). However, while important, this criticism raises issues orthogonal to the ones at stake here, and will therefore not be discussed further.

2. This understanding of evolutionary psychology is thus closer to what Buller (2005) calls "ep," rather than "EP."

3. Note that the extent to which evolved traits ought to be expected to be species-universal is a current issue of debate: see, e.g., Buss and Hawley (2010).

4. For example, I will not consider the criticism that evolutionary psychologists fail to pay sufficient attention to individual differences in the expression of a given trait (see, e.g., Schulz, 2010; but see also Buss & Hawley, 2010).

5. A further issue that needs to be addressed here is that the "trait" in question has to in fact be a trait in its own right (and not merely a part of a set of other traits); for more on this worry, see Dupre (2002); Gould and Lewontin (1979); Richardson (2007, chap. 2). In general, in order to deal with this worry, the only plausible approach is to consider whether taking the "trait" in question to be a genuine trait makes for productive research. In what follows, I presume this to be the case. It is also worth noting that, as made clearer in chapter 2, this definitely seems to be true when it comes to representational decision making.

6. These conditions are unified by the fact that, conjointly, they try to identify the factors that can impact the evolution of a given trait.

7. For another reason to be doubtful about the distinction between exaptation and adaptation, see Schulz (2013b).

8. This variability needs to be seen to include any developmental constraints on the trait in question: a trait that cannot develop in certain ways has a low degree of effective variability.

9. Note that this worry mirrors some of the concerns with "adaptationism" in evolutionary biology more generally (a point further confirmed by the fact that

Richardson's 2007 discussion of evolutionary psychology relies heavily on Brandon's 1990 discussion of adaptationist reasoning in evolutionary biology; for more on adaptationism, also see Orzack & Sober, 1994; and Godfrey-Smith, 2001).

10. Note that I do not put these alternative hypotheses forward as credible contenders; I use them merely for expository purposes.

11. Note that I do not want to defend an epistemic separability assumption for all cases of (causal) explanation: the claim is just that it is plausible when it comes to evolutionary biological analyses. (This thus goes against some of the remarks made in Brandon, 1990; see also Godfrey-Smith, 2009.).

12. Another way to put this is that the difference between what Brandon (1990) calls "how actually" explanations (which satisfy all of the above conditions) and "how possibly" explanations (which do not) is one of degree, not kind. For a different defense of the robust epistemic value of "how possibly" explanations, see Forber (2010). For an example of a partial evolutionary biological analysis that is widely seen as epistemically plausible, see the work on sex ratio theory (Charnov et al., 1981; Trivers & Willard, 1973; Sober, 2010).

13. I move away from the case of the food preferences for salty and fatty foods here, as the situation with regards to behavioral innovation is a little clearer. While this is not a case of evolutionary *psychology* per se—the focus is on dispositions to innovate behaviors, without regards to the psychological causes of these behaviors—the methodological upshot here is the same.

14. Note also that empirically determining when behavioral innovation occurs, while not trivial, is not impossibly complex (Lefebvre & Bolhuis, 2003).

15. This claim about when and why behavioral innovation is adaptive can also be used as evidence for other claims—such as the idea that behavioral innovation should be expected to be unequally distributed among the sexes (Lee, 2003; Reader & Laland, 2003). See also below.

16. Saying exactly what this means is tricky, due to the fact that many models are idealized. However, for present purposes, this point does not need to be settled. For more on this, see, e.g., Weisberg (2007a); Cartwright (1999); Hausman (1992).

17. For example: should outliers and ceiling effects be excluded from the analysis? How are the data to be coded? There are often multiple plausible answers to these questions.

Chapter 4

1. Here, it is also important to keep in mind that some of the work that *seems* to be about the evolution of representational decision making turns out to actually be about related, but importantly different issues. For example, there is some work in

the literature in economics on the evolution of "preferences": see, e.g., Guth (1995); Robson and Samuelson (2008); Samuelson (2001); Robson (2001). However, the aim of this work is quite different from the approach taken in this book. For example, Robson (2001) tries to determine when motivational structures that allow for learning are more adaptive than ones that are "hardwired"—which, though, cross-cuts the issues of importance here. Similarly, Samuelson (2001) tries to determine when complexity constraints lead organisms to move away from a purely associationist motivational architecture; his argument too, though, is based on the idea that associationist architectures must be innate and unchangeable. Much the same goes for Sober (1994a), which concerns the evolution of different sorts of cognitive representational decision making, and not the evolution of representational decision making per se. Finally, Kerr (2007) presents a model of how an organism's interactions with its environment can change its behavioral dispositions—however, again, this is not about the evolution of representational decision making per se. I return to some of these issues below.

2. Millikan spells this out by appealing to Gibson's work on ecological psychology (see, e.g., Gibson, 1979; Reed, 1996). For present purposes, though, this is not so important.

3. Here and in what follows, I use "representational decision making" as shorthand for "decision making based on distinct cognitive and conative representations."

4. A fourth potential problem of Millikan's account concerns the fact that, if anything, it is more plausible that conative representational decision making evolved before cognitive representational decision making—and not the other way around, as seems to be implied in her account (see, e.g., Schroeder et al., 2010). However, since the issues here are somewhat complex—see chapter 6 for more on this—I will not consider this worry for Millikan's account further here.

5. There are many differences between the accounts of the authors mentioned in the text, but for present purposes, it sufficient to focus on their communalities.

6. Indeed, as Papineau (2003, pp. 104–106) notes, through classical conditioning, they can even make some general, causal inferences—albeit in an "egocentric" way that is centered on the organism's own actions.

7. The rest of this section draws on parts of Schulz (2011a) and Schulz (2013a).

8. It thus seems that Sterelny (2003) does not distinguish between the cognitive representation that one is dehydrated and the conative representation to find water. While there is room to question this conflation (see also Spurrett, 2015; Christensen, 2010), I will not do so here. See chapter 6 for more on this.

9. Sterelny attributes this point to Dickinson: see Sterelny (2003, p. 93).

10. Further criticisms are found in Sober (1997) and Walsh (1997). However, the issues they raise are quite different from the ones brought up here: in particular, they

are more concerned with the complexity thesis, whereas I am more concerned with the flexibility thesis.

11. A structurally related point has been made by Block (1981) in a different context. I thank Kim Sterelny for useful remarks about this matter.

12. It could be that Sterelny here (and in relation to the first alleged benefit of preferences) reasons as follows: a preference-driven organism could just rely on a few "ultimate" preferences, and use these in combination with its decoupled representations to determine the appropriate behavior for the case at hand. For example, the organism could just have a preference for fighting (say), and then use its decoupled representations to determine the appropriate way to fight in the situation it is in (see Sterelny, 2003, pp. 87–90). If so, then his account would come closer to mine in chapter 6. However, it remains true that a more detailed picture of this kind of reasoning—and of the benefits it brings—is necessary to make this account plausible.

13. For example, an organism that is motivated to consume both carbohydrates and proteins, but who is more in need of the former than the latter, could just spend more time foraging for carbohydrates than for protein (in a ratio determined by the ratio of the strength of the two needs).

14. Even some neuroscientific accounts of decision making employ a principle of this sort: see, e.g., Glimcher et al. (2005).

15. Sterelny (in personal conversation) has claimed that the fact that preferences have structured content can also be used to answer this question (see also note 8). However, it is not clear why this would be so: if, for different reasons, I both prefer p to q and q to p, it is not clear how appealing to the details of what p and q are can help me decide what to do.

16. This is less plausible for instrumental preferences—however, as Sterelny (2003, pp. 87–90) himself notes, it is not so clear that we want to rely on these in this context. Note also that the fact that drives are acquired quite quickly is also important in the context of economic arguments for the evolution of preferences—see chapter 3, note 14.

17. On top of this, it is important to keep in mind that, as noted above, the possibility for causal reasoning needs to be seen as another benefit of representational decision making, at least in some circumstances. Any novel account needs to be consistent with this. However, in practice, this should not be so hard to satisfy (and is true for the account defended in this book).

Chapter 5

1. An early expression of some of the ideas behind this account is in Schulz (2011a). Note that this account (and the one in the next chapter) employs a cost-benefit

analysis approach toward the evolution of representational decision making. For more on such cost-benefit analyses, see Lee (2003) and Rosenzweig (1996). Note also that somewhat similar considerations have been appealed to by other authors (see, e.g., Sober, 1981; Lycan, 1988; Bennett, 1964, 1976, 1991; see also Stich, 1990; de Sousa, 2007), but for very different—largely epistemological—purposes, and without spelling them out in the way done here. See also note 9 below.

2. Note that this is not to say that robust tracking should be expected to evolve in all cases. The point is just that it is plausible that it will often do so (see also Sterelny, 2003, chaps. 2–3). See also below.

3. As made clearer in chapter 4, Millikan (2002) makes a similar point as well. Also see Whiten (1995).

4. Here, it is also important to note that my point is not that non-representational decision makers need to track the world with very finely individuated perceptual states. My point is that, whatever the perceptual states are like that the organism relies on, cognitive representations allow for their grouping—and thus, for streamlining the decision-making process.

5. This chunking may also depend on innate expectations about likely sensory inputs.

6. For some of the classic work on neural pruning, see Changeux et al. (1973); Edelman and Moutcastle (1978); Edelman (1978). Note also that, while there is still some controversy surrounding the importance of neural pruning in neural development, the latter has come to be increasingly widely recognized as a key element of neural ontogeny (at least in the way it is understood here); see the quotes and references in the text.

7. An implication of this is that neuronal selection proceeds non-linearly across a person's life: see Gazzaniga et al. (2009) and E. Santos and Noggle (2011).

8. Note that the point here is not that organisms *could not* have an inefficient decision-making system. The point is just that it is adaptive for an organism to make efficient use of its biological resources: for even if it is possible for it to develop and use large decision-making systems, this does not mean that it is adaptive to do so. After all, energy is a scarce resource for all organisms, and thus there is reason to expect its neurobiological systems to be efficiently built.

9. Note also that this confirms the point made in chapter 2 (see note 14) that not all parts of cognitive science ought to be expected to make much appeal to cognitive representations.

10. Much the same goes for many kinds of close-contact interactions: fights with conspecifics, mating behaviors, play, locomotion, and so on. See also Clark (1997).

11. See also Carruthers (2006) and Millikan (2002) for more on this.

12. There is also another reason why cognitive representations are often modeled as being graded. This reason centers on the fact that non-graded cognitive representations seem epistemically foundationalist. This may be thought to be problematic, as it seems that there is always the possibility that whatever mechanism the organism uses to ground its cognitive representations has misfired for one reason or another. Hence, it might seem plausible that, to the extent that they are epistemically rational, agents instead should allow for graded cognitive representations (see also Joyce, 1998, 2009). However, since none of this actually speaks to what is in fact going on in the minds of real organisms, but instead evaluates what it would be epistemically *rational* for ideal agents to be like, this reason for seeing cognitive representations as graded is not so important in the present context.

13. In fact, it might (at least in principle) also be possible to extend my account to allow for genuinely graded cognitive representations (see, e.g., D. W. Stephens, 1989). However, it is likely that this would require a radical departure from the basic foundations of the picture here set out. The reason for this is that the core idea of this picture is that perceptual cue-determined action *already incorporates* an "expected utility" calculation of sorts: perceptual cues need not be fully reliable to adaptively trigger a certain action; as long as they more often than not lead to the adaptive action, always triggering that action can be adaptive (depending on the availability of more reliable cues; Godfrey-Smith, 1991, 1996; Cooper, 2001; D. W. Stephens, 1989). This point naturally carries over to my account of cognitive representational decision making: it need not always be fully reliable to form a given cognitive representation in the presence of a given perceptual cue—as long as it is more reliable than chance, relying on the perceptual cue can be adaptive. For this reason, allowing for genuinely graded cognitive representations is not greatly plausible on my account, and I will not discuss this possibility further here. At any rate, as is made clearer in the text, there is no need to appeal to genuinely graded cognitive representations, as there are other ways of handling the work based on them.

14. Note also that it is possible to extend this account to non-probabilistic measures of graded cognitive representations.

15. It is worthwhile to note, though, that the translation from graded accounts of cognitive representations into the above two proposals can be somewhat complex. So, for example, it is conceivable that one wants to model an organism that is certain to degree 0.7 that the probability that the state of the world is S is 0.3. Translating this into the second proposal above requires seeing the organism as representing the fact that it is representing the fact that the state of the world is S with probability 0.3 with probability 0.7—that is, it requires seeing this organism as being meta-representational. However, this does not fundamentally change the point made in the text.

16. For the same reason, my account is also consistent with all of the major accounts of how cognitive representations should be seen to be changed across time—some of

which appeal to graded cognitive representations (see, e.g., Bradley, 2005) and some of which do not (see, e.g., Alchourrón et al., 1985).

17. A somewhat related issue concerns the question of whether cognitive representations should be expected to be true (for discussion, see, e.g., Stich, 1990; Godfrey-Smith, 1991, 1996; C. L. Stephens, 2001; Sober, 1994b, 1981; Lycan, 1988). However, my account is consistent with any way this debate turns out: for this debate turns on the relationship between the truth of the content of a cognitive representation and the adaptiveness of acting on that cognitive representation. My account, though, only assumes that acting on the relevant cognitive representations is (sufficiently often) adaptive—whether or not their content is descriptively accurate is left open.

18. Moreover, my account of the evolution of conative representational decision making is entirely consistent with the modularity of mind as well: see chapters 6 and 8. For a defense of the modularity of mind that is built on a notion of cognitive efficiency (though one that differs somewhat from the one at stake here), see also Sperber (2005).

Chapter 6

1. For an earlier version of this account, see Schulz (2013). A related neuroscientific account is presented in Koscik & Tranel (2012), though the latter places more emphasis on the evolutionary importance of social environments than is done here.

2. A closely related realistic case may be bee flight: see, e.g., Montague et al. (1995).

3. Note that, as will be made clearer below, table 6.1 really summarizes two different decisions the organism is making: where to move to in the west-east dimension, and where to move to in the south-north dimension.

4. Note that this may require that the organism cognitively represents the grid itself. However, this is just a feature of this particular example—as made clearer in chapter 2, organisms can be pure conative representational decision makers, without needing to be cognitive representational decision makers at all.

5. As was true in chapter 5, the exact details of these network representations do not matter here—the networks are just intended to make the argument in the text clearer.

6. For a more realistic example of this sort of case, see Hasselmo (2012, pp. 90–97).

7. As also noted in chapter 5, another set of costs sometimes appealed to here concerns the fact that representational decision makers seem to be faced with non-graceful degradation in function: if they are damaged, they are unable to function at all, whereas non-representational decision makers can still function partially even if somewhat damaged (see, e.g., Rumelhart et al., 1986). However, as also noted in chapter 5, this point is misleading, as it will depend on the kind of damage at issue.

Damage to a non-representational decision maker that prevents it from accessing its memory store appropriately can also lead to catastrophic failures in functioning, and damage to the inferential abilities of a conative representational decision maker might lead to it being a slower or less accurate conative representational decision maker, but still a functional one. Because of this, I do not consider the question of the graceful or non-graceful degradation in function of conative representational decision makers further here.

8. Note that this rule is used just as an illustrative example; it is likely that the real rule used would be more complex (see, e.g., Mitani & Watts, 2001; Boesch, 2002; Hawkes & Bird, 2002; Barrett, 2005).

9. Similar remarks can be made about other causal environments—such as ones surrounding different kinds of tool use (see also Sterelny, 2003, pp. 45–50).

10. Again, the rule is used here merely as an illustrative example.

11. Note that this example does not concern mindreading. That is, my account does not predict that conative representational decision making would be only (or especially) adaptive in cases where mindreading is adaptive. (See also Andrews, 2012; and Sterelny, 2003, chap. 4, for accounts of why the ability to mindread others may not be necessary to navigate even complex social environments.).

12. I return to this point in chapter 7.

13. It is important to note that many apparent cases of intransitive choices may not, in fact, be such cases, since the relevant choice options may have been misdescribed or the organism may have changed its behavioral goals (see, e.g., Bratman, 1987; Broome, 1991, 1999; A. I. Houston et al., 2007; Johnson & Busemeyer, 2005; Hagen et al., 2012; Kahneman & Tversky, 1979; Grether & Plott, 1979; Kalenscher et al., 2010). Still, it does seem plausible that, even after these corrections have been made, many cases of intransitive choices remain (see also Johnson & Busemeyer, 2005; Cox & Epstein, 1989; Loomes et al., 1991).

14. For a different argument for the conclusion that humans (at least) should be expected to frequently rely on non-monotonic conative representations, see Schulz (2015b).

15. Some conative representations may be unavailable as possible drivers of the organism's behavior, as they compute over cognitive representations (say) that are differently structured than the ones that the organism is currently tokening. To return to the above example: the organism may be in a situation in which there are no hunting parties to join, and hence cannot make the conative representation "join any hunt for which $0.5 <$ [size of the prey animal (in pounds) / size of hunting group] < 1.5; in case of several options, choose the first one offered" the basis of its actions. Still, it is plausible that there will often be several other conative representations remaining that are possible drivers of the organism's behavior.

16. Related remarks can be made concerning conative representations for impossible goals (e.g., that I discover the largest prime number, or that I will win the lottery). In general—keeping in mind the constraint noted in chapter 2 that conative representations need to be able to be acted on—such impossible goals do not pose particular problems for my account, as even impossible goals can lead to specific behavioral outcomes (the goal to prove that there is a largest prime might get me to engage in mathematical studies, for example; see also Wall, 2009). In fact, the resultant behaviors can even be adaptive (my mathematical endeavors—even though they are ultimately mathematically futile—might make me highly attractive to potential mates, for example).

17. That is: while conative and cognitive representational decision making are adaptive in the same broad types of environments, they are not necessarily adaptive in all the same sub-types of these environments.

18. Another example that might be cited here concerns bee navigation. As noted by Sterelny (2003, chaps. 2–3), bees that are forced to return to their nest when they cannot see the nest or any other landmarks simply recall the direction they were flying in before, and continue accordingly.

19. One complication worth noting here is that, as also made clearer in chapter 2, some of the neural foundations of the conative representational decision-making system seem to overlap with those of the non-representational decision-making system (probably due to the fact that both systems have to result in action somehow). This is makes it less than fully clear whether it is really true that conative representational decision making, as it is understood here, really evolved before cognitive representational decision making. Fortunately, for present purposes, this issue does not need to be resolved.

20. With some notable exceptions—e.g., that of the nematode worm *C. elegans*.

Chapter 7

1. There is also much work suggesting that cognition should be seen to be "embodied"—that is, to involve the organism's body in key ways (Chemero, 2009; Clark, 1997; Shapiro, 2004, 2011a; Alsmith & de Vignemont, 2012). While some of what I say in this chapter also applies to this issue, the focus will be on the (possible) extendedness of representational decision making.

2. Of course, it is possible to doubt that much of cognition is representational for reasons that have nothing to do with the EMT (Hutto & Myin, 2012; Gallagher, 2015). It is a goal of the rest of this book to build part of the case against this view.

3. As with the AEP argument of chapter 3, this argument is not as such presented in the literature. However, various authors state arguments close to this: see, e.g., Brooks (1991); van Gelder (1995); Rowlands (1999); Beer (1990); L. B. Smith and Thelen

(1994); Silberstein and Chemero (2012). At any rate, I take it that this argument is interesting independently of who has put it forward.

4. Note that what follows in the rest of this section differs from the arguments in Shapiro (2011b), in that the latter argues that evolutionary theoretic considerations do not give support to the idea that cognition is *extended* (or even embedded). By contrast, the goal in this section is to show that evolutionary theoretic considerations do not support the idea that cognition is *non-representational*. In fact, as the next section will go on to show, I think that there *is* evolutionary theoretic support for the idea that cognition is sometimes extended (or embedded; though this support relies on different considerations from those criticized by Shapiro, 2011b). Note, finally, that Shapiro (2011b) also expresses doubt about the general plausibility of evolutionary biological arguments for the nature and reality of a given cognitive architecture; however, these concerns have already been addressed in chapter 3 above.

5. Note that another way of lightening these cognitive loads is by moving toward decision rules that are relatively simple and can thus be quickly and easily applied, but which are still quite accurate (see, e.g., Gigerenzer & Selten, 2001). Discussing these further is the aim of the next chapter (but it is worthwhile to note here that it is possible to see the discussion of this chapter as a special case of that in the next chapter: externalized representational decision making employs specialist decision rules of a specific kind—namely, ones that depend, in a specific way, on the external environment).

6. Clark and Toribio (1994) also present arguments for seeing cognition as involving both representational and non-representational elements. However, their arguments are quite different from the ones set out here.

7. I here assume that the organism is a cognitive representational decision maker, but it is possible to extend the argument that follows to *purely* conative representational (i.e., non-cognitive representational) decision makers as well.

8. Of course, it is possible that this way of externalizing decision making has other benefits (such as increased decision-making accuracy). However, it is not clear that the existence of these benefits is a necessary or general feature of this way of making decisions, so I will not consider this further here.

9. For more on altruistic decision making, see chapter 9 below.

10. Note that there are two main ways in which such a division of labor can be realized—in line with the two "radical" ways of externalizing representational decision making laid out above. First, organisms can specialize in different decision-making domains, with some organisms focusing on making decisions in one set of circumstances, and others in other sets of circumstances. For example, organism A could specialize in foraging for berries—e.g., by deciding where to search for (edible) berries largely reflexively—and organism B could specialize in finding suitable camping

grounds—e.g., by deciding where to look for suitable camping grounds largely reflexively. Then, organism A could rely on the decision rule "do whatever organism B does" when looking for suitable camping grounds, and organism B could rely on the decision rule "do whatever organism A does" when foraging for berries. The benefit of this would be that both organisms can make both foraging and camping decisions nearly as fast and accurately as a purely reflexively driven organism—but with only half the cognitive requirements of the latter. The second way in which cooperative social organisms could divide the labor of representational decision making is by computing different parts of the same decision problem. So, as in the above case, one organism could calculate the food-recovery rate of the current patch, and the other organism could calculate the average food-recovery rate of the other patches. They could then exchange this information, just relying on the decision rule "determine the food-recovery rate of the current patch (or the other patches), and compare with what organism B (or A) says; stay if and only if the former is higher than the latter." In this way, they would gain all of the benefits of a purely representationally driven organism—but could make decisions more quickly and with less recourse to concentration and attention. For present purposes, though, a closer discussion of these different ways of dividing the labor of representational decision making is not necessary.

11. Someone might object that the above has only shown that human representational decision making will sometimes be embedded—not that it will also be extended: there might be further conditions that need to be satisfied for it to be genuinely extended (Clark, 1997; Clark & Chalmers, 1998). In response, I note, on the one hand, that the exact nature of these further conditions is controversial (see also Shapiro, 2011b), and on the other, that there is no reason these further conditions would not also be satisfied in the above cases.

Chapter 8

1. The details of these approaches are complex, and differ among the different versions of RCT: see, e.g., Savage (1954), Jeffrey (1983), and Joyce (1999).

2. For more on purely behaviorist and normative approaches to RCT, see, e.g., Hausman (2012) and Bermudez (2009). Further, note that Gigerenzer et al. also reject the idea that "classical" (RCT-based, probabilistic, and logical) standards of rationality are normatively compelling (Brighton & Todd, 2009; Gigerenzer & Selten, 2001; Gigerenzer, 2007, 2008). However, since this book is not concerned with questions about "rationality," I will not discuss this further here—though see also the concluding section of this chapter.

3. Note that this does not mean that the implementation of these generalist, optimizing decision rules does not involve some sort of generalist heuristics, like Markov chain Monte Carlo analysis (Colombo & Hartmann, 2017). The point is that the aim is to determine the optimal behavioral response to the situation the organism is in, taking as much of the available information into account as possible.

4. In this, they take their cue from the work of Herbert Simon (see, especially, Simon, 1957), but elaborate his theory along several dimensions. For parallels in the context of animal decision making, see, e.g., Chater (2012); Hagen et al. (2012).

5. Note that a special case of such a simple heuristic is a rule that says "do what it says on this sheet" (or something of this sort). The particular benefits and disadvantages of this simple heuristic are discussed in more detail in chapter 7. See also Brighton and Todd (2009).

6. Simon (1957) also argues that the relevant thresholds should be seen as differing from situation to situation. However, this is not so important here.

7. In fact, Gigerenzer et al. argue that, in a certain sense, simple heuristics can be even more accurate than optimal decision rules: for example, when the problem involves estimation of an unknown quantity, the best statistical method (e.g., a form of Bayesian inference) can be less reliable than certain simple heuristics (see, e.g., Gigerenzer & Selten, 2001). However, for present purposes, this complication can be left aside—all that matters here is that the success with which simple heuristics solve decision problems depends on features of the environment; whether this success is slightly higher, slightly lower, or about equal to that of optimal decision rules is secondary here.

8. Something similar has also be shown for many animals: see, e.g., Trimmer et al. (2008).

9. Note that Gigerenzer et al. do not claim the considerations they put forward are the only ones that are relevant for the evolution of simple heuristics-based decision making (and, as noted in the previous section, these are not the only considerations they put forward in support of their theory). However, they do suggest that there is evolutionary biological support for their theory.

10. For further criticisms of these arguments, see Schulz (2011b).

11. See also Cosmides and Tooby (1987), Tooby and Cosmides (1992), and Tooby et al. (2005) for a—quite different—argument for a somewhat similar conclusion.

12. As noted in chapters 5 and 6, I here leave open how the relevant conative representations are acquired. In the present context, this is important, as it is entirely possible that an organism starts with a general conative representation, and then, with time and practice, comes to acquire more specialized conative representations. This is consistent with my account in this chapter, as the latter only concerns the conditions under which acting on specialist vs. generalist conative representations is adaptive—not the manner in which these are acquired.

13. The latter is a major issue that is much discussed in the literature on simple heuristics (see, e.g., Kruglanski & Gigerenzer, 2011). Note also that some decision rules might, in virtue of their nature, only be applicable to certain circumstances (e.g., "when your child needs help, provide the help" only applies to decision situations

that concern helping one's child; see the next chapter for more on this). However, in general, the organism needs to be provided with information as to which decision rule is to be used (or to be used first) in which circumstances.

14. This is of course not to say that the environment *cannot* change in a way that affects all of the relevant decisions problems equally: for example, it is possible that, due to climate change, all foraging decisions need to be altered in similar ways, as all foodstuffs start to ripen more quickly. The point is just that it is relatively unlikely that this happens on a regular basis (at least if compared to uncorrelated changes in these environments).

15. At this point, someone might object that according to my account, fully generalist conative representational decision making does not evolve. While my account can allow for differently general simple heuristics, it does not make room for cases where an organism makes all decisions by relying on only one generalist decision rule (such as "maximize expected utility"). However, I am happy to accept this: for the reasons laid out in the text, I think that it is just not plausible that an organism would rely solely on a fully general conative representational decision rule. See also the next chapter for further reasons in support of this conclusion.

16. Gigerenzer et al. do not deny that these other kinds of problems are key examples of their approach as well (see, e.g., Kruglanski & Gigerenzer, 2011). The point made in the text is just that the example of the baseball player is easily read as suggesting that all fast motor decisions are good examples of the use of simple heuristics—something that I think is false.

Chapter 9

1. This chapter is an edited and much revised version of Schulz (2016a).

2. Note that it is possible to formulate more drastically different understandings of these notions (see, e.g., Garson, 2016; Piccinini & Schulz, forthcoming). However, this is not greatly problematic, for while the arguments of this chapter would need to be reformulated if a drastically different understanding of psychological altruism were used, the core ideas of this chapter would remain substantively the same (though they would then not concern psychological altruism per se, but some related notion). Note also that psychological altruism as defined here must not be conflated with *evolutionary altruism*: organismic traits that provide (relative or absolute) fitness benefits to other organisms (Sober & Wilson, 1998; Okasha, 2006). The latter raises different issues from the ones at stake here, and will not be discussed further in what follows. In line with this, all unqualified references to "altruism" or "egoism" in this chapter should be taken to refer to the psychological varieties of these two theses only.

3. As in the previous chapter, I leave open whether these behavioral dispositions are triggered by cognitive representations or by perceptual cues.

4. Note that the argument of this section also shows that other generalist helping architectures—that is, generalist helping architectures that are not egoistic (such as ones focused on Hamilton's rule)—can be evolutionarily unstable. This thus further reinforces some of the points made in the previous chapter. See also note 7 below.

5. Though it is important to note that it is possible that different human populations faced sufficiently different environmental conditions and that only some found altruistic conative representations toward their offspring to be adaptive. For present purposes, though, it is enough to note that it is plausible that at least some human populations are among those that have found altruism to be adaptive.

6. As made clear in note 13, there are possible exceptions to this point—which, however, do not invalidate its importance here.

7. Note that this argument also shows why it is plausible that relying on other general conative representations—such as Hamilton's rule—is, for many organisms, not adaptive (for more on Hamilton's rule, see, e.g., Gardner et al., 2011). Organisms that assess whether $rB > C$ every time their offspring is in need are likely to be less fit than organisms that just presume that, for offspring, this inequality is generally satisfied: the benefits from saving the computational costs are likely to be greater than the costs of occasional errors. I return to this point briefly in the concluding section. That said, it is entirely possible that a conative representation to help all close relatives (not just offspring) is adaptive in some contexts (Gardner et al., 2011; A. S. Griffin & West, 2002). However, for present purposes, focusing on the case of helping offspring is sufficient.

8. Moreover, since the content of this conative representation contains its own area of application—namely, decisions concerning whether to help offspring in need—there is little information that needs to be stored about *when* being an altruist is adaptive.

9. I agree with Sober and Wilson (1998) that there is no reason to think that psychological altruism is any more or less energetically costly than psychological egoism (Sober & Wilson, 1998, p. 322). My point is just the altruist can make decisions relatively more quickly and with less recourse to concentration and attention than the egoist does. (Of course, it is possible that, at least sometimes, the cognitive efficiency of an organism can be indirectly measured by its caloric intake—a more cognitively efficient organism might have more time foraging and consuming food, for example. However, it would then remain the case that the more cognitively efficient mind design might itself be just as energy hungry as the less cognitively efficient one—it is just that the former can cause the organism to lower its energy consumption overall. In other words: it may be that developing and maintaining traits T and T' is energetically equally costly, but that, once either T or T' is in place, T leads the organism to lower its overall energy consumption relative to T'. I thank Justin Garson for useful discussion of this point.)

10. Note also that exactly the same point could be made concerning the conative representation to help one's offspring, which might be seen as in fact targeted at particular kinds of behaviors—i.e., offspring-helping ones. The same point as the one made in the text applies here, too.

11. In principle, this point could also arise for offspring-directed (non-reciprocal) help. So for example, for many mammals (including humans), shielding infant offspring with their own bodies from physical dangers is generally adaptive: as noted earlier, infant offspring bodies are often much less robust than adult bodies, and can thus suffer great damage from even minor collisions. Since a conative representation instructing organisms to shield their infant offspring with their own body is very close to a motor command already, it is thus reasonable to think that shielding infants in a non-conative representational way is adaptive. However, this case seems less clear cut than the reciprocation-based one in the text, so I focus on the latter one here.

12. Male vampire bats also seem to reciprocate by grooming (Carter & Wilkinson, 2013). However, this may be a separate case of reciprocation. At any rate, this does not affect the substance of the point made in the text.

13. The parenthetical remark is important: it is entirely possible that some organisms—such as humans—evolved the disposition to form and act on morality-based conative representations (e.g., concerning equality or respect for personal liberties) as well. However, this calls for an investigation of its own.

14. Put differently: it is an implication of the arguments of this chapter that human helping behaviors (for one) form a heterogeneous class, and should be recognized as such.

References

Abernethy, B., & Russell, D. G. (1987). The relationship between expertise and visual search strategy in a racquet sport. *Human Movement Science, 6*(4), 283–319.

Adams, F., & Aizawa, K. (2008). *The Bounds of Cognition*. Oxford: Blackwell.

Aizawa, K. (2003). *The Systematicity Argument*. Dordrecht: Kluwer.

Alchourrón, C. E., Gärdenfors, P., & Makinson, D. (1985). On the logic of theory change: Partial meet contraction and revision functions. *Journal of Symbolic Logic, 50*, 510–530.

Alcock, J. (2013). *Animal Behavior: An Evolutionary Approach* (10th ed.). Sunderland: Sinauer Associates.

Allen, C. (1992). Mental content. *British Journal for the Philosophy of Science, 43*(4), 537–553.

Allen, C. (1999). Animal concepts revisited: The use of self-monitoring as an empirical approach. *Erkenntnis, 51*(1), 537–544.

Allen, C. (2004). Is anyone a cognitive ethologist? *Biology & Philosophy, 19*, 589–607.

Allen, C. (2014). Models, mechanisms, and animal minds. *Southern Journal of Philosophy, 52*, 75–97.

Allen, C., & Bekoff, M. (1994). Intentionality, social play, and definition. *Biology & Philosophy, 9*(1), 63–74.

Allen, C., & Bekoff, M. (1997). *Species of Mind: The Philosophy and Biology of Cognitive Ethology*. Cambridge, MA: MIT Press.

Allen, C., Bekoff, M., & Lauder, G. V. (Eds.). (1998). *Nature's Purposes: Analyses of Function and Design in Biology*. Cambridge, MA: MIT Press.

Allen, C., & Hauser, M. (1991). Concept attribution in nonhuman animals: Theoretical and methodological problems in ascribing complex mental processes. *Philosophy of Science, 58*, 221–240.

Alsmith, A., & de Vignemont, F. (2012). Embodying the mind and representing the body. *Review of Philosophy and Psychology, 3*(1), 1–13.

Anderson, C. W. (1993). The modulation of feeding behavior in response to prey type in the frog Rana pipiens. *Journal of Experimental Biology, 179*(1), 1–12.

Anderson, J. R. (1993). Problem solving and learning. *American Psychologist, 48*(1), 35–44.

Anderson, J. R. (2007). *How Can the Mind Occur in a Physical Universe.* Oxford: Oxford University Press.

Anderson, J. R., Bothell, D., Lebiere, C., & Matessa, M. (1998). An integrated theory of list memory. *Journal of Memory and Language, 38,* 341–380.

Andrews, K. (2012). *Do Apes Read Minds? Toward a New Folk Psychology.* Cambridge, MA: MIT Press.

Andrews, K. (2015). *The Animal Mind. An Introduction to the Philosophy of Animal Cognition.* London: Routledge.

Anscombe, E. (2000). *Intention* (2nd ed.). Cambridge, MA: Harvard University Press.

Ariew, A., Cummins, R., & Perlman, M. (Eds.). (2002). *Functions: New Essays in the Philosophy of Psychology and Biology.* Oxford: Oxford University Press.

Auletta, G. (2011). *Cognitive Biology: Dealing with Information from Bacteria to Minds.* Oxford: Oxford University Press.

Bago, B., & De Neys, W. (2017). Fast logic? Examining the time course assumption of dual process theory. *Cognition, 158,* 90–109.

Baillargeon, R., Scott, R. M., & He, Z. (2010). False-belief understanding in infants. *Trends in Cognitive Sciences, 14*(3), 110–118.

Barkow, J., Cosmides, L., & Tooby, J. (Eds.). (1992). *The Adapted Mind.* Oxford: Oxford University Press.

Barrett, H. C. (2005). Adaptations to predators and prey. In D. M. Buss (Ed.), *The Handbook of Evolutionary Psychology* (pp. 200–223). Hoboken, NJ: Wiley.

Barrett, H. C. (2015). *The Shape of Thought: How Mental Adaptations Evolve.* Oxford: Oxford University Press.

Barrows, E. M. (2011). *Animal Behavior Desk Reference: A Dictionary of Animal Behavior, Ecology, and Evolution* (3rd ed.). Boca Raton: Taylor & Francis.

Barsalou, L. W. (1999). Perceptual symbol systems. *Behavioral and Brain Sciences, 22*(4), 577–609; discussion 610–560.

Batson, D. (1991). *The Altruism Question: Toward a Social-Psychological Answer.* Hillsdale, NJ: Lawrence Erlbaum Associates.

Baum, D. A., & Larson, A. (1991). Adaptation reviewed: A phylogenetic methodology for studying character macroevolution. *Systematic Zoology, 40*, 1–18.

Baum, D. A., & Smith, S. D. (2013). Tree *Thinking:* An Introduction to Phylogenetic Biology. Greenwood Village, CO: Roberts & Company.

Beer, R. D. (1990). *Intelligence as Adaptive Behavior.* New York: Academic Press.

Bennett, J. (1964). *Rationality.* London: Keegan & Paul.

Bennett, J. (1976). *Linguistic Behavior.* Cambridge: Cambridge University Press.

Bennett, J. (1991). How to read minds in behavior: A suggestion from a philosopher. In A. Whiten (Ed.), *Natural Theories of Mind: Evolution, Development and Simulation of Everyday Mind Reading* (pp. 97–108). Oxford: Blackwell Publishing.

Berkman, E., & Lieberman, M. D. (2009). The neuroscience of goal pursuit: Bridging gaps between theory and data. In M. G. Grant & H. Grant (Eds.), *The Psychology of Goals* (pp. 98–126). New York: Guilford Press.

Bermudez, J. (2009). *Decision Theory and Rationality.* Oxford: Oxford University Press.

Binmore, K. (1998). Game Theory and the Social Contract (Vol. II): *Just Playing.* Cambridge, MA: MIT Press.

Blakemore, R. (1975). Magnetotactic bacteria. *Science, 190*(4212), 377–379.

Block, N. (1981). Psychologism and behaviorism. *Philosophical Review, 90*, 5–43.

Boesch, C. (1994). Cooperative hunting in wild chimpanzees. *Animal Behaviour, 48*(3), 653–667.

Boesch, C. (2002). Cooperative hunting roles among taï chimpanzees. *Human Nature, 13*(1), 27–46.

Borensztajn, G., Zuidema, W., & Bechtel, W. (2014). Systematicity and the need for encapsulated representations. In P. Calvo & J. Symons (Eds.), *The Architecture of Cognition* (pp. 165–189). Cambridge, MA: MIT Press.

Boyd, R., & Richerson, P. (1985). *Culture and the Evolutionary Process.* Chicago: University of Chicago Press.

Boyd, R., & Richerson, P. (2005). *The Origin and Evolution of Cultures.* Oxford: Oxford University Press.

Boyd, R., Richerson, P., & Henrich, J. (2011). The cultural niche: Why social learning is essential for human adaptation. *Proceedings of the National Academy of Sciences of the United States of America, 108*(Supplement 2), 10918–10925.

Bradley, R. (2005). Radical probabilism and Bayesian conditioning. *Philosophy of Science, 72*(2), 342–364.

Bradley, R., & List, C. (2009). Desire-as-belief revisited. *Analysis, 69*(1), 31–37.

Brandon, R. (1990). *Adaptation and Environment*. Princeton: Princeton University Press.

Bratman, M. (1987). *Intention, Plans, and Practical Reason*. Cambridge, MA: Harvard University Press.

Brewin, C. R. (1989). Cognitive change processes in psychotherapy. *Psychological Review, 96*, 379–394.

Brighton, H. J., & Todd, P. M. (2009). Situating rationality: Ecologically rational decision making with simple heuristics. In P. A. Robbins & M. Aydede (Eds.), *The Cambridge Handbook of Situated Cognition* (pp. 322–346). Cambridge: Cambridge University Press.

Brooks, R. (1991). Intelligence without representation. *Artificial Intelligence, 47*, 139–159.

Broome, J. (1991). *Weighing Goods*. Oxford: Blackwell.

Broome, J. (1999). *Ethics out of Economics*. Cambridge: Cambridge University Press.

Brown, G., Wood, A., & Chater, N. (2012). Sources of variation within the individual. In P. Hammerstein & J. R. Stevens (Eds.), *Evolution and the Mechanisms of Decision Making* (pp. 227–242). Cambridge, MA: MIT Press.

Brown, J. W., & Braver, T. S. (2005). Learned predictions of error likelihood in the anterior cingulate cortex. *Science, 307*(5712), 1118–1121.

Buck, R. (1991). Motivation, emotion and cognition: A developmental-interactionist view. In K. T. Strongman (Ed.), *International Review of Studies on Emotion* (Vol. 1, pp. 101–142). New York: Wiley.

Buhl, J., Deneubourg, J. L., Grimal, A., & Theraulaz, G. (2005). Self-organized digging activity in ant colonies. *Behavioral Ecology and Sociobiology, 58*(1), 9–17.

Buller, D. (2005). *Adapting Minds*. Cambridge, MA: MIT Press.

Burge, T. (2010). *Origins of Objectivity*. Oxford: Oxford University Press.

Buss, D. M. (2014). *Evolutionary Psychology: The New Science of the Mind* (5th ed.). Boston: Allyn & Bacon.

Buss, D. M., & Hawley, P. H. (Eds.). (2010). *The Evolution of Personality and Individual Differences*. Oxford: Oxford University Press.

Buss, D. M., & Schmitt, D. P. (1993). Sexual strategies theory: An evolutionary perspective on human mating. *Psychological Review, 100*(2), 204–232.

Byrne, R. W. (2003). Novelty in deceit. In S. M. Reader & K. N. Laland (Eds.), *Animal Innovation* (pp. 237–259). Oxford: Oxford University Press.

Calvo, P., Martin, E., & Symons, J. (2014). The emergence of systematicity in minimally cognitive agents. In P. Calvo & J. Symons (Eds.), *The Architecture of Cognition* (pp. 397–434). Cambridge, MA: The MIT Press.

Calvo, P., & Symons, J. (Eds.). (2014). *The Architecture of Cognition.* Cambridge, MA: MIT Press.

Camp, E. (2004). The generality constraint and categorical restrictions. *Philosophical Quarterly, 54*(215), 209–231.

Camp, E. (2009). Putting thoughts to work. *Philosophy and Phenomenological Research, 78,* 275–311.

Campbell, R., & Kumar, V. (2012). Moral reasoning on the ground. *Ethics, 122*(2), 273–312.

Carey, S. (2011). *The Origin of Concepts.* Oxford: Oxford University Press.

Carey, S., & Spelke, E. (1996). Science and core knowledge. *Philosophy of Science, 63,* 515–533.

Carruthers, P. (2002). Review of "Human Nature and the Limits of Science" by John Dupré. *Economics and Philosophy, 18,* 357–363.

Carruthers, P. (2006). *The Architecture of the Mind.* Oxford: Oxford University Press.

Carter, G., & Wilkinson, G. (2013). Food sharing in vampire bats: Reciprocal help predicts donations more than relatedness or harassment. *Proceedings. Biological Sciences, 280,* 20122573.

Cartwright, N. (1999). *The Dappled World: A Study of the Boundaries of Science.* Cambridge: Cambridge University Press.

Casey, B. J., Giedd, J. N., & Thomas, K. M. (2000). Structural and functional brain development and its relation to cognitive development. *Biological Psychology, 54*(1–3), 241–257.

Casey, B. J., Thomas, K. M., Davidson, M. C., Kunz, K., & Franzen, P. L. (2002). Dissociating striatal and hippocampal function developmentally with a stimulus–response compatibility task. *Journal of Neuroscience, 22*(19), 8647–8652.

Chaiken, S., & Trope, Y. (Eds.). (1999). *Dual-Process Theories in Social Psychology.* New York: The Guilford Press.

Changeux, J.-P., Courrége, P., & Danchin, A. (1973). A theory of the epigenesis of neuronal networks by selective stabilization of synapses. *Proceedings of the National Academy of Sciences of the United States of America, 70*(10), 2974–2978.

Charnov, E. L., Los-den Hartogh, R. L., Jones, W. T., & van den Assem, J. (1981). Sex ratio evolution in a variable environment. *Nature, 289,* 27–33.

Chater, N. (2012). Building blocks of human decision making. In P. Hammerstein & J. R. Stevens (Eds.), *Evolution and the Mechanisms of Decision Making* (pp. 53–68). Cambridge, MA: MIT Press.

Chemero, A. (2009). *Radical Embodied Cognitive Science*. Cambridge, MA: MIT Press.

Cheng, P. W., & Holyoak, K. J. (1985). Pragmatic reasoning schemas. *Cognitive Psychology, 17*(4), 391–416.

Cherniak, C. (1986). *Minimal Rationality*. Cambridge, MA: MIT Press.

Cherniak, C. (2012). Neural wiring optimization. In M. Hofman & D. Falk (Eds.), *Evolution of the Primate Brain: From Neuron to Behavior* (pp. 361–372). Amsterdam: Elsevier.

Chevrund, J., & Wolf, J. (2009). The genetics and evolutionary consequences of maternal effects. In D. Maestripieri & J. Mateo (Eds.), *Maternal Effects in Mammals* (pp. 11–37). Chicago: University of Chicago Press.

Christensen, W. (2010). The decoupled representation theory of the evolution of cognition—A critical assessment. *British Journal for the Philosophy of Science, 61*, 361–405.

Churchland, P. (1985). *Neurophilosophy*. Cambridge, MA: MIT Press.

Churchland, P. (2011). *Braintrust: What Neuroscience Tells Us about Morality*. Princeton: Princeton University Press.

Cialdini, R. B., Brown, S. L., Lewis, B. P., Luce, C., & Neuberg, S. L. (1997). Reinterpreting the empathy-altruism relationship: When one into one equals oneness. *Journal of Personality and Social Psychology, 73*(3), 481–494.

Cirulli, F., Berry, A., & Alleva, E. (2003). Early disruption of the mother-infant relationship: Effects on brain plasticity and implications for psychopathology. *Neuroscience and Biobehavioral Reviews, 27*(1–2), 73–82.

Clark, A. (1997). *Being There*. Cambridge, MA: MIT Press.

Clark, A. (2008). *Supersizing the Mind*. Oxford: Oxford University Press.

Clark, A. (2013). Whatever next: Predictive brains, situated agents, and the future of cognitive science. *Behavioral and Brain Sciences, 36*(3), 1–73.

Clark, A., & Chalmers, D. (1998). The extended mind. *Analysis, 58*, 7–19.

Clark, A., & Toribio, J. (1994). Doing without representing. *Synthese, 101*(3), 401–431.

Clavien, C., & Chapuisat, M. (2013). Altruism across disciplines: One word, multiple meanings. *Biology & Philosophy, 28*, 125–140.

Clutton-Brock, T. (2009). Cooperation between non-kin in animal societies. *Nature, 462*(7269), 51–57.

Colombo, M. (2013). Moving forward (and beyond) the modularity debate: A network perspective. *Philosophy of Science, 80*, 356–377.

Colombo, M. (2014). Explaining social norm compliance. A plea for neural representations. *Phenomenology and the Cognitive Sciences, 13*(2), 217–238.

Colombo, M. (2017). Social motivation in computational neuroscience. Or, if brains are prediction machines, then the Humean theory of motivation is false. In J. Kiverstein (Ed.), *Routledge Handbook of Philosophy of the Social Mind* (pp. 320–340). London: Routledge.

Colombo, M., & Hartmann, S. (2017). Bayesian cognitive science, unification, and explanation. *British Journal for the Philosophy of Science, 68*, 451–484.

Constantinescu, A. O., O'Reilly, J. X., & Behrens, T. E. J. (2016). Organizing conceptual knowledge in humans with a gridlike code. *Science, 352*(6292), 1464–1468.

Cooper, W. S. (2001). *The Evolution of Reason: Logic as a Branch of Biology.* Cambridge University Press.

Cosmides, L., & Tooby, J. (1987). From evolution to behavior: Evolutionary psychology as the missing link. In J. Dupre (Ed.), *The Latest on the Best: Essays on Evolution and Optimality* (pp. 277–306). Cambridge, MA: MIT Press.

Cosmides, L., & Tooby, J. (1992). Cognitive adaptations for social exchange. In J. Barkow, L. Cosmides, & J. Tooby (Eds.), *The Adapted Mind: Evolutionary Psychology and the Generation of Culture* (pp. 163–228). Oxford: Oxford University Press.

Cosmides, L., & Tooby, J. (2008). Can a general deontic logic capture the facts of human moral reasoning? How the mind interprets social exchange rules and detects cheaters. In W. Sinnott-Armstrong (Ed.), *Moral Psychology* (Vol. I—The Evolution of Morality: Adaptations and Innateness, pp. 53–119). Cambridge, MA: MIT Press.

Cox, J. C., & Epstein, S. (1989). Preference reversals without the independence axiom. *American Economic Review, 79*(3), 408–426.

Crane, T. (2016). *The Mechanical Mind: A Philosophical Introduction to Minds, Machines and Mental Representation* (3rd ed.). London: Routledge.

Cummins, R. (1996). *Representations, Targets, and Attitudes.* Cambridge, MA: MIT Press.

Curley, J., & Keverne, E. (2005). Genes, brains and mammalian social bonds. *Trends in Ecology & Evolution, 20*, 561–567.

D'Amato, M. R., & van Sant, P. (1988). The person concept in monkeys (Cebus apella). *Journal of Experimental Psychology. Animal Behavior Processes, 14*(1), 43–55.

Damasio, A. (1994). *Descartes' Error: Emotion, Reason, and the Human Brain.* New York: Avon Books.

Davidson, D. (1980). *Essays on Actions and Events*. Oxford: Oxford University Press.

Davidson, D. (1982). Rational animals. *Dialectica, 36*, 318–327.

Dawkins, R. (1986). *The Blind Watchmaker*. New York: Norton.

Dayan, P. (2012). Robust neural decision making. In P. Hammerstein & J. R. Stevens (Eds.), *Evolution and the Mechanisms of Decision Making* (pp. 151–167). Cambridge, MA: MIT Press.

De Finetti, B. (1980). Foresight: Its logical laws, its subjective sources. In H. E., Kyburg Jr. & H. E. Smokler (Eds.), *Studies in Subjective Probability* (2nd ed., pp. 93–158). Huntington, NY: Krieger.

De Hevia, M. D., Izard, V., Coubart, A., Spelke, E. S., & Streri, A. (2014). Representations of space, time, and number in neonates. *Proceedings of the National Academy of Sciences of the United States of America, 111*, 4809–4813.

De Sousa, R. (2007). *Why Think? Evolution and the Rational Mind*. Oxford: Oxford University Press.

Deban, S. M., O'Reilly, J. C., & Nishikawa, K. C. (2001). The evolution of the motor control of feeding in amphibians. *American Zoologist, 41*(6), 1280–1298.

Dekel, E., & Scotchmer, S. (1999). On the evolution of attitudes towards risk in winner-take-all games. *Journal of Economic Theory, 87*, 125–143.

Dempster, A. P. (1968). A generalization of Bayesian inference. *Journal of the Royal Statistical Society. Series B. Methodological, 30*, 205–247.

Dennett, D. C. (1987). *The Intentional Stance*. Oxford: Blackwell.

Di Paolo, E. A. (2005). Autopoiesis, adaptivity, teleology, agency. *Phenomenology and the Cognitive Sciences, 4*, 97–125.

Dial, K. P., Randall, R. J., & Dial, T. R. (2006). What use is half a wing in the ecology and evolution of birds? *Bioscience, 56*(5), 437–445.

Diaz-Uriarte, R., & Garland, J. T. (1996). Testing hypotheses of correlated evolution using phylogenetically independent contrasts: Sensitivity to deviations from Brownian motion. *Systematic Biology, 45*(1), 27–47.

Dickinson, A. (1985). Actions and habits: The development of behavioural autonomy. *Proceedings of the Royal Society of London. Series B, Biological Sciences, 308*(1135), 67–78.

Dickinson, A., & Balleine, B. (2000). Causal cognition and goal-directed action. In C. Heyes & L. Huber (Eds.), *The Evolution of Cognition* (pp. 185–204). Cambridge, MA: MIT Press.

Dickinson, A., Balleine, B., Watt, A., Gonzales, F., & Boakes, R. A. (1995). Motivational control after extended instrumental training. *Animal Learning & Behavior, 23*(2), 197–206.

Dolan, R. J., & Dayan, P. (2013). Goals and habits in the brain. *Neuron, 80*(2), 312–325.

Dovidio, J., Piliavin, J., Schroeder, D., & Penner, L. (2006). *The Social Psychology of Prosocial Behavior*. Mahawan, N.J.: Lawrence Erlbaum Associates.

Dretske, F. (1981). *Knowledge and the Flow of Information*. Cambridge, MA: MIT Press.

Dretske, F. (1988). *Explaining Behavior*. Cambridge, MA: MIT Press.

Dupre, J. (1996). The mental lives of nonhuman animals. In C. Allen & D. Jamison (Eds.), *Readings in animal cognition* (pp. 323–336). Cambridge, MA: MIT Press.

Dupre, J. (2002). Ontology is the problem. *Behavioral and Brain Sciences, 25*, 516–517.

Earman, J. (1992). *Bayes or Bust? A Critical Examination of Bayesian Confirmation Theory*. Cambridge, MA: MIT Press.

Edelman, G. M. (1978). *Neural Darwinism: The Theory of Neuronal Group Selection*. Oxford: Oxford University Press.

Edelman, G. M., & Moutcastle, V. B. (1978). *The Mindful Brain*. Cambridge, MA: MIT Press.

Epstein, S. (1994). Integration of the cognitive and the psychodynamic unconscious. *American Psychologist, 49*(8), 709–724.

Epstein, S., Lipson, A., Holstein, C., & Huh, E. (1992). Irrational reactions to negative outcomes: Evidence for two conceptual systems. *Journal of Personality and Social Psychology, 62*, 328–339.

Ericsson, K. A., & Charness, N. (1994). Expert performance: Its structure and acquisition. *American Psychologist, 49*(8), 725–747.

Evans, G. (1982). *The Varieties of Reference*. Oxford: Clarendon Press.

Evans, J. S. (2008). Dual-processing accounts of reasoning, judgment, and social cognition. *Annual Review of Psychology, 59*, 255–278.

Evans, J. S., & Frankish, K. (Eds.). (2009). *In Two Minds: Dual Process and Beyond*. Oxford: Oxford University Press.

Faivre, D., & Schüler, D. (2008). Magnetotactic bacteria and magnetosomes. *Chemical Reviews, 108*(11), 4875–4898.

Falk, A., Fehr, E., & Fischbacher, U. (2003). On the nature of fair behavior. *Economic Inquiry, 41*(1), 20–26.

Faure, A., Haberland, U., Condé, F., & Massioui, N. E. (2005). Lesion to the nigrostriatal dopamine system disrupts stimulus-response habit formation. *Journal of Neuroscience*, *25*(11), 2771–2780.

Fehr, C. (2012). Feminist engagement with evolutionary psychology. *Hypatia*, *27*(1), 50–72.

Fehr, E., & Camerer, C. F. (2007). Social neuroeconomics: The neural circuitry of social preferences. *Trends in Cognitive Sciences*, *11*(10), 419–427.

Fehr, E., & Gaechter, S. (2000). Fairness and retaliation: The economics of reciprocity. *Journal of Economic Perspectives*, *14*, 159–181.

Felleman, D. J., & van Essen, D. C. (1991). Distributed hierarchical processing in the primate cerebral cortex. *Cerebral Cortex*, *1*, 1–47.

Felsenstein, J. (2004). *Inferring Phylogenies*. Sunderland: Sinauer Associates.

Fidgor, C. (2011). Semantics and metaphysics in informatics: Towards an ontology of tasks. *Topics in Cognitive Science*, *3*, 222–226.

Firestone, C., & Scholl, B. J. (2016). Cognition does not affect perception: Evaluating the evidence for "top-down" effects. *Behavioral and Brain Sciences*, *39*, e229.

Fischman, M. G., & Schneider, T. (1985). Skill level, vision, and proprioception in simple one-hand catching. *Journal of Motor Behavior*, *17*(2), 219–229.

Fisher, J. A. (1996). The myth of anthropomorphism. In M. Bekhoff & D. Jamieson (Eds.), *Readings in Animal Cognition* (pp. 3–16). Cambridge, MA: MIT Press.

Fitzgibbon, C. D. (1995). Comparative ecology of two elephant-shrew species in a Kenyan coastal forest. *Mammal Review*, *25*(1–2), 19–30.

Fodor, J. (1983). *The Modularity of Mind: An Essay on Faculty Psychology*. Cambridge, MA: MIT Press.

Fodor, J. (1990). *The Theory of Content*. Cambridge, MA: MIT Press.

Fodor, J. (2008). Comment on Cosmides & Tooby. In W. Sinnott-Armstrong (Ed.), *Moral Psychology* (Vol. I—The Evolution of Morality: Adaptations and Innateness, pp. 137–141). Cambridge, MA: MIT Press.

Fodor, J., & Pylyshyn, Z. (1988). Connectionism and cognitive architecture: A critical analysis. *Cognition*, *28*, 3–71.

Foley, R. (1992). The epistemology of belief and the epistemology of degrees of belief. *American Philosophical Quarterly*, *29*(2), 111–121.

Forber, P. (2010). Confirmation and explaining how possible. *Studies in the History and Philosophy of Biological and Biomedical Sciences*, *41*, 32–40.

Franco, L., & Cannas, S. (1998). Solving arithmetic networks using feed-forward neural networks. *Neurocomputing, 18*, 61–79.

Galef, B. G. (2012). Social learning and traditions in animals: Evidence, definitions, and relationship to human culture. *Wiley Interdisciplinary Reviews: Cognitive Science, 3*(6), 581–592.

Galef, B. G., & Laland, K. N. (2005). Social learning in animals: Empirical studies and theoretical models. *Bioscience, 55*(6), 489–499.

Gallagher, S. (2015). Re-presenting representation. *Philosophical Inquiries, 3*, 71–84.

Gallistel, C. (1990). *The Organization of Learning*. Cambridge, MA: MIT Press.

Gallistel, C. (2000). The replacement of general-purpose learning models with adaptively specialized learning modules. In M. Gazzaniga (Ed.), *The Cognitive Neurosciences* (2nd ed., pp. 1179–1191). Cambridge, MA: MIT Press.

Gallistel, C., & King, A. (2009). *Memory and the Computational Brain*. Oxford: Wiley-Blackwell.

Gardner, A., West, S. A., & Wild, G. (2011). The genetical theory of kin selection. *Journal of Evolutionary Biology, 24*(5), 1020–1043.

Garnier, S., Gautrais, J., & Theraulaz, G. (2007). The biological principles of swarm intelligence. *Swarm Intelligence, 1*(1), 3–31.

Garson, J. (2014). *The Biological Mind*. London: Routledge.

Garson, J. (2016). Two types of psychological hedonism. *Studies in History and Philosophy of Science Part C: Studies in History and Philosophy of Biological and Biomedical Sciences, 56*, 7–14.

Gazzaniga, M., Ivry, R. B., & Mangum, G. R. (2009). *Cognitive neuroscience: The biology of the mind* (3rd ed.). New York: W.W. Norton & Company.

Gendler, T. (2008). Alief and belief. *Journal of Philosophy, 105*, 634–663.

Gibson, J. J. (1966). *The Senses Considered as Perceptual Systems*. Boston: Houghton Mifflin.

Gibson, J. J. (1979). *The Ecological Approach to Visual Perception*. Boston: Houghton Mifflin.

Gigerenzer, G. (2007). *Gut Feelings: The Intelligence of the Unconscious*. New York: Penguin.

Gigerenzer, G. (2008). *Rationality for Mortals*. Oxford: Oxford University Press.

Gigerenzer, G., & Selten, R. (Eds.). (2001). *Bounded Rationality: The Adaptive Toolbox*. Cambridge, MA: MIT Press.

Gillespie, J. (1998). *Population Genetics: A Concise Guide* (2nd ed.). Baltimore: Johns Hopkins University Press.

Glimcher, P. W., Dorris, M. C., & Bayer, H. M. (2005). Physiological utility theory and the neuroeconomics of choice. *Games and Economic Behavior, 52*(2), 213–256.

Godfrey-Smith, P. (1991). Signal, decision, action. *Journal of Philosophy, 88*(12), 709–722.

Godfrey-Smith, P. (1996). *Complexity and the Function of Mind in Nature.* Cambridge: Cambridge University Press.

Godfrey-Smith, P. (2001). Three kinds of adaptationism. In S. H. Orzack & E. Sober (Eds.), *Adaptationism and Optimality* (pp. 335–357). Cambridge: Cambridge University Press.

Godfrey-Smith, P. (2002). On the evolution of representational and interpretive capacities. *Monist, 85,* 50–69.

Godfrey-Smith, P. (2009). *Darwinian Populations and Natural Selection.* Oxford: Oxford University Press.

Goel, V., & Dolan, R. J. (2003). Explaining modulation of reasoning by belief. *Cognition, 87*(1), B11–B22.

Goldman, A. (1970). *A Theory of Human Action.* Princeton: Princeton University Press.

Goldman, A. (2006). *Simulating Minds.* Oxford: Oxford University Press.

Gooding, D., Pinch, T., & Schaffer, S. (1989). *The Uses of Experiment: Studies in the Natural Sciences.* Cambridge: Cambridge University Press.

Goodman, S. N. (1999a). Toward evidence-based medical statistics. 1: The P value fallacy. *Annals of Internal Medicine, 130,* 995–1004.

Goodman, S. N. (1999b). Toward evidence-based medical statistics. 2: The Bayes factor. *Annals of Internal Medicine, 130,* 1005–1013.

Goodman, S. N., & Royall, R. (1988). Evidence and scientific research. *American Journal of Public Health, 78*(12), 1568–1574.

Gopnik, A., Glymour, C., Sobel, D. M., Schulz, L. E., Kushnir, T., & Danks, D. (2004). A theory of causal learning in children: Causal maps and Bayes nets. *Psychological Review, 111*(1), 3–32.

Gopnik, A., & Schulz, L. E. (2004). Mechanisms of theory-formation in young children. *Trends in Cognitive Sciences, 8*(8), 371–377.

Gould, S. J., & Lewontin, R. (1979). The Spandrels of San Marco and the Panglossian paradigm: A critique of the adaptationist programme. *Proceedings of the Royal Society of London. Series B, Biological Sciences, 205*(1161), 581–598.

Gould, S. J., & Vrba, E. (1982). Exaptation—A missing term in the science of form. *Paleobiology, 8*, 4–15.

Grabner, R. H., Fink, A., Stipacek, A., Neuper, C., & Neubauer, A. C. (2004). Intelligence and working memory systems: Evidence of neural efficiency in alpha band ERD. *Brain Research. Cognitive Brain Research, 20*(2), 212–225.

Grau, J. (2002). Learning and memory without a brain. In M. Bekhoff, C. Allen, & G. Burghardt (Eds.), *The Cognitive Animal* (pp. 77–88). Cambridge, MA: MIT Press.

Graybiel, A. (2008). Habits, rituals, and the evaluative brain. *Annual Review of Neuroscience, 31*, 359–387.

Greene, J. (2008). The secret joke of Kant's Soul. In W. Sinnott-Armstrong (Ed.), *Moral Psychology* (Vol. 3, pp. 35–79). Cambridge, MA: MIT Press.

Grether, D., & Plott, C. (1979). Economic theory of choice and the preference reversal phenomenon. *American Economic Review, 69*, 623–638.

Griffin, A. S., & West, S. A. (2002). Kin selection: Fact and fiction. *Trends in Ecology & Evolution, 17*, 15–21.

Griffin, D. R. (1984). *Animal Thinking*. Cambridge, MA: Harvard University Press.

Griffin, S. R., Smith, M. L., & Seeley, T. D. (2012). Do honeybees use the directional information in round dances to find nearby food sources? *Animal Behaviour, 83*, 1319–1324.

Griffiths, P., & Stotz, K. (2000). How the mind grows. *Synthese, 122*, 29–51.

Grimm, D. (2014). In dogs' play, researchers see honesty and deceit, perhaps something like morality. *Washington Post*, May 19. http://www.washingtonpost.com /national/health-science/in-dogs-play-researchers-see-honesty-and-deceit-perhaps -something-like-morality/2014/05/19/d8367214-ccb3-11e3-95f7-7ecdde72d2ea _story.html.

Guala, F. (2005). *The Methodology of Experimental Economics*. Cambridge: Cambridge University Press.

Guth, W. (1995). An evolutionary approach to explaining cooperative behavior by reciprocal incentives. *International Journal of Game Theory, 24*, 323–344.

Hagen, E., Chater, N., Gallistel, C., Houston, A. I., Kacelnik, A., & Kalenscher, T., Nettle, D., Oppenheimer, D., & Stephens, D. (2012). Decision making: What can evolution do for us? In P. Hammerstein & J. R. Stevens (Eds.), *Evolution and the Mechanisms of Decision Making* (pp. 97–126). Cambridge, MA: MIT Press.

Haidt, J. (2001). The emotional dog and its rational tail: A social intuitionist approach to moral judgment. *Psychological Review, 108*, 814–834.

Haidt, J., & Kesebir, S. (2010). Morality. In S. Fiske, D. Gilbert, & G. Lindzey (Eds.), *Handbook of Social Psychology* (5th ed., pp. 797–832). Hoboken, NJ: Wiley.

Halford, G. S., Wilson, W. H., & Phillips, S. (1998). Processing capacity defined by relational complexity: Implications for comparative, developmental, and cognitive psychology. *Behavioral and Brain Sciences, 21*(6), 803–831.

Hammerstein, P. (2003). Why is reciprocity so rare in social animals? A Protestant appeal. In P. Hammerstein (Ed.), *Genetic and Cultural Evolution of Cooperation* (pp. 83–94). Cambridge, MA: MIT Press.

Hammerstein, P., & Stevens, J. R. (Eds.). (2012). *Evolution and the Mechanisms of Decision Making.* Cambridge, MA: MIT Press.

Hare, B., Call, J., Agnetta, B., & Tomasello, M. (2000). Chimpanzees know what conspecifics do and do not see. *Animal Behaviour, 59,* 771–786.

Harré, R. (2002). Social reality and the myth of social structure. *European Journal of Social Theory, 5*(1), 111–123.

Harvey, P. H., Read, A. F., & Nee, S. (1995a). Further remarks on the role of phylogeny in comparative ecology. *Journal of Ecology, 83*(4), 733–734.

Harvey, P. H., Read, A. F., & Nee, S. (1995b). Why ecologists need to be phylogenetically challenged. *Journal of Ecology, 83*(3), 535–536.

Hasselmo, M. (2012). *How We Remember.* Cambridge, MA: MIT Press.

Haugeland, J. (1999). Mind embodied and embedded. In J. Haugeland (Ed.), *Having Thought* (pp. 207–237). Cambridge, MA: Harvard University Press.

Hausfater, G., & Hrdy, S. B. (Eds.). (1984). *Infanticide: Comparative and Evolutionary Perspectives.* Chicago: Aldine Transactions.

Hausman, D. M. (1992). *The Inexact and Separate Science of Economics.* Cambridge: Cambridge University Press.

Hausman, D. M. (2012). *Preference, Value, Choice, and Welfare.* Cambridge: Cambridge University Press.

Hawkes, K., & Bird, R. B. (2002). Showing off, handicap signaling, and the evolution of men's work. *Evolutionary Anthropology, 11,* 58–67.

Hawthorne, J. (2009). The Lockean thesis and the logic of belief. In F. Huber & C. Schmidt-Petri (Eds.), *Degrees of Belief* (Vol. 342, pp. 49–74). Dordrecht: Springer.

Henrich, J. (2015). *The Secret of Our Success: How Culture Is Driving Human Evolution, Domesticating Our Species, and Making Us Smarter.* Princeton, NJ: Princeton University Press.

Heyes, C. M. (2012). Grist and mills: On the cultural origins of cultural learning. *Phil. Trans. R. Soc. B, 367,* 2181–2191.

Heyes, C. M. (2013). Imitation—associative and context-dependent. In W. Prinz, M. Beisert, & A. Herwig (Eds.), *Action Science: Foundations of an Emerging Discipline* (pp. 309–332). Cambridge, MA: MIT Press.

Heyes, C. M. (2014a). False belief in infancy: A fresh look. *Developmental Science, 17*(5), 647–659.

Heyes, C. M. (2014b). Rich interpretations of infant behaviour are popular but are they valid? *Developmental Science, 17,* 665–666.

Heyes, C. M., & Frith, C. (2014). The cultural evolution of mind reading. *Science, 344,* 1243091.

Hobbs, K., & Spelke, E. S. (2015). Goal attributions and instrumental helping at 14 and 24 months of age. *Cognition, 142,* 44–59.

Houston, A. I., & McNamara, J. M. (1999). *Models of Adaptive Behaviour: An Approach Based on State.* Cambridge: Cambridge University Press.

Houston, A. I., McNamara, J. M., & Steer, M. (2007). Violations of transitivity under fitness maximization. *Biology Letters, 3,* 365–367.

Howson, C., & Urbach, P. (2006). *Scientific Reasoning: The Bayesian Approach* (3rd ed.). Peru: Open Court.

Hrdy, S. B. (1999). *Mother Nature.* New York: Ballantine.

Huber, F. (2009). Belief and degrees of belief. In F. Huber & C. Schmidt-Petri (Eds.), *Degrees of Belief* (pp. 1–33). Dordrecht: Springer.

Humphrey, N. (1986). *The Inner Eye: Social Intelligence in Evolution.* Oxford: Oxford University Press.

Humphreys, P. (2004). *Extending Ourselves: Computational Science, Empiricism, and Scientific Method.* Oxford: Oxford University Press.

Hurley, S. L. (2005). Social heuristics that make us smarter. *Philosophical Psychology, 18*(5), 585–612.

Hutto, D. D., & Myin, E. (2012). *Radicalizing Enactivism: Basic Minds without Content.* Cambridge, MA: MIT Press.

Jamieson, D., & Bekoff, M. (1992). On aims and methods of cognitive ethology. PSA: Proceedings of the Biennial Meeting of the Philosophy of Science Association, 110–124.

Jamieson, I. G., & Craig, J. L. (1987). Critique of helping behavior in birds: A departure from functional explanations. In P. P. G. Bateson & P. Klopfer (Eds.), (pp. 79–98). *Perspectives in Ethology.* New York: Springer US.

Jarvstad, A., Hahn, U., Rushton, S. K., & Warren, P. A. (2013). Perceptuo-motor, cognitive, and description-based decision-making seem equally good. *Proceedings of the National Academy of Sciences of the United States of America, 110*(40), 16271–16276.

Jarvstad, A., Hahn, U., Warren, P. A., & Rushton, S. K. (2014). Are perceptuo-motor decisions really more optimal than cognitive decisions? *Cognition, 130*(3), 397–416.

Jeffrey, R. (1983). *The Logic of Decision* (2nd ed.). Chicago: University of Chicago Press.

Jeffrey, R. (1992). *Probability and the Art of Judgment.* Cambridge: Cambridge University Press.

Jensen, K. (2012). Who cares? Other-regarding concerns—Decisions with feeling. In P. Hammerstein & J. R. Stevens (Eds.), *Evolution and the Mechanisms of Decision Making* (pp. 299–317). Cambridge, MA: MIT Press.

Johnson, J. G., & Busemeyer, J. R. (2005). A dynamic, stochastic, computational model of preference reversal phenomena. *Psychological Review, 112*(4), 841–861.

Joyce, J. M. (1998). A nonpragmatic vindication of probabilism. *Philosophy of Science, 65*, 575–603.

Joyce, J. M. (1999). *The Foundations of Causal Decision Theory.* Cambridge: Cambridge University Press.

Joyce, J. M. (2009). Accuracy and coherence: Prospects for an alethic epistemology of partial belief. In F. Huber & C. Schmidt-Petri (Eds.), *Degrees of Belief* (Vol. 342, pp. 263–297). Dordrecht: Springer.

Joyce, J. M. (2010). A defense of imprecise credences in inference and decision making. *Philosophical Perspectives, 24*, 281–323.

Kacelnik, A. (2012). Putting mechanisms into behavioral ecology. In P. Hammerstein & J. R. Stevens (Eds.), *Evolution and the Mechanisms of Decision Making* (pp. 21–38). Cambridge, MA: MIT Press.

Kahneman, D. (2003). Maps of bounded rationality: Psychology for behavioral economics. *American Economic Review, 93*, 1449–1475.

Kahneman, D., & Tversky, A. (1979). Prospect theory. *Econometrica, 47*, 263–291.

Kalenscher, T., Tobler, P., Huijbers, W., Daselaar, S., & Pennartz, C. (2010). Neural signatures of intransitive preferences. *Frontiers in Human Neuroscience, 4*, 1–14.

Kareev, Y. (2012). Advantages of cognitive limitations. In P. Hammerstein & J. R. Stevens (Eds.), *Evolution and the Mechanisms of Decision Making* (pp. 169–182). Cambridge, MA: MIT Press.

Katz, L. D. (2005). Review of Timothy Schroeder: "Three Faces of Desire." *Notre Dame Philosophical Reviews.* https://ndpr.nd.edu/news/24858-three-faces-of-desire/

Kerr, B. (2007). The evolution of cognition. *Biological Theory, 2,* 250–262.

Kim, H. F., & Hikosaka, O. (2015). Parallel basal ganglia circuits for voluntary and automatic behaviour to reach rewards. *Brain, 138,* 1776–1800.

Kirsh, D. (1996). Today the earwig, tomorrow man? In M. Boden (Ed.), *The Philosophy of Artificial Life* (pp. 237–261). Oxford: Oxford University Press.

Kitcher, P. (1985). *Vaulting Ambition: Sociobiology and the Quest for Human Nature.* Cambridge, MA: MIT Press.

Kitcher, P. (2011). *The Ethical Project.* Cambridge, MA: Harvard University Press.

Klaes, C., Westendorff, S., Chakrabarti, S., & Gail, A. (2011). Choosing goals, not rules: Deciding among rule-based action plans. *Neuron, 70*(3), 536–548.

Klein, S. B., Cosmides, L., Tooby, J., & Chance, S. (2002). Decisions and the evolution of memory: Multiple systems, multiple functions. *Psychological Review, 109*(2), 306–329.

Koenig, W. D., & Mumme, R. L. (1990). Levels of analysis and the functional significance of helping behavior. In M. Bekoff & D. Jamieson (Eds.), *Interpretation and Explanation in the Study of Behavior: Explanation, Evolution, and Adaptation* (Vol. 2, pp. 268–303). Boulder, CO: Westview Press.

Korf, R. E. (1987). Planning as search: A quantitative approach. *Artificial Intelligence, 33*(1), 65–88.

Koscik, T. R., & Tranel, D. (2012). Brain evolution and human neuropsychology: The inferential brain hypothesis. *Journal of the International Neuropsychological Society, 18*(3), 394–401.

Krantz, D. H. (1972). A theory of magnitude estimation and cross-modality matching. *Journal of Mathematical Psychology, 9*(2), 168–199.

Kripke, S. (1982). *Wittgenstein on Rules and Private Language.* Cambridge, MA: Harvard University Press.

Kruglanski, A. W., & Gigerenzer, G. (2011). Intuitive and deliberate judgments are based on common principles. *Psychological Review, 118*(1), 97–109.

Kuzdzal-Fick, J. A., Fox, S. A., Strassmann, J. E., & Queller, D. C. (2011). High relatedness is necessary and sufficient to maintain multicellularity in Dictyostelium. *Science, 334,* 1548–1551.

Laland, K. N., & van Bergen, Y. (2003). Experimental studies of innovation in the guppy. In S. M. Reader & K. N. Laland (Eds.), *Animal Innovation* (pp. 155–174). Oxford: Oxford University Press.

Langen, T. A. (1999). How western scrub-jays (Alphelocoma californica) select a nut: Effects of the number of options, variation in nut size, and social competition among foragers. *Animal Cognition, 2,* 223–233.

Lauder, G. V., & Reilly, S. M. (1994). Amphibian feeding behavior: Comparative biomechanics and evolution. In V. Bels, M. Chardon, & P. Vandewalle (Eds.), *Biomechanics of Feeding in Vertebrates* (Vol. 18, pp. 163–195). Springer Berlin Heidelberg.

Lee, P. C. (2003). Innovation as a behavioral response to environmental challenges: A cost and benefit approach. In S. M. Reader & K. N. Laland (Eds.), *Animal Innovation* (pp. 261–278). Oxford: Oxford University Press.

Lefebvre, L., & Bolhuis, J. J. (2003). Positive and negative correlates of feeding innovations in birds: Evidence for limited modularity. In S. M. Reader & K. N. Laland (Eds.), *Animal Innovation* (pp. 39–62). Oxford: Oxford University Press.

Leventhal, H. (1982). The integration of emotion and cognition: A view from the perceptual-motor theory of emotion. In M. S. Clark & S. T. Fiske (Eds.), *Affect and cognition: The Seventeenth Annual Carnegie Symposium on Cognition* (pp. 121–156). Hillsdale, NJ: Lawrence Erlbaum Associates.

Levy, W. B., & Steward, O. (1979). Synapses as associative memory elements in the hippocampal formation. *Brain Research, 175*(2), 233–245.

Lewontin, R. (1970). The units of selection. *Annual Review of Ecology and Systematics, 1*, 1–18.

Lewontin, R. (1998). The evolution of cognition: Questions we will never answer. In D. Scarborough & S. Sternberg (Eds.), *Methods, Models, and Conceptual Issues: An Invitation to Cognitive Science* (Vol. 4, pp. 107–132). Cambridge, MA: MIT Press.

Lieberman, M. D. (2003). Reflexive and reflective judgment processes: A social cognitive neuroscience approach. In J. P. Forgas, K. D. Williams, & W. von Hippel (Eds.), *Social Judgments: Implicit and Explicit Processes* (pp. 44–67). New York, NY: Cambridge University Press.

Lieberman, M. D. (2010). Social cognitive neuroscience. In S. T. Fiske, D. T. Gilbert, & G. Lindzey (Eds.), *Handbook of Social Psychology* (5th ed., pp. 143–193). New York: McGraw Hill.

Linn, I., Perrin Michael, R., & Hiscocks, K. (2007). Use of space by the four-toed elephant-shrew Petrodromus tetradactylus (Macroscelidae) in Kwazulu-Natal (South Africa). *Mammalia, 71*, 30.

Loomes, G., Starmer, C., & Sugden, R. (1991). Observing violations of transitivity by experimental methods. *Econometrica, 59*, 425–439.

Loomes, G., & Sugden, R. (1982). Regret theory: An alternative theory of choice under uncertainty. *Economic Journal (London), 92*(368), 805–824.

Lycan, W. (1988). *Judgement and Justification*. Cambridge: Cambridge University Press.

Lyons, J. C. (2001). Carving the mind at its (not necessarily modular) joints. *British Journal for the Philosophy of Science, 52*, 277–302.

Machery, E. (2009). *Doing without Concepts*. Oxford: Oxford University Press.

Machery, E. (forthcoming). Discovery and confirmation in evolutionary psychology. In J. Prinz (Ed.), *Oxford Handbook of Philosophy of Psychology*. Oxford: Oxford University Press.

Machery, E., & Barrett, H. C. (2006). Debunking adapting minds. *Philosophy of Science, 73*, 232–246.

Mackintosh, N. J. (1983). *Conditioning and Associative Learning*. Oxford: Clarendon Press.

Mackintosh, N. J. (1994). *Animal Learning and Conditioning* (2nd ed.). San Diego: Academic Press.

Marcus, G. (2014). PDP and symbol manipulation: What's been learned since 1986? In P. Calvo & J. Symons (Eds.), *The Architecture of Cognition* (pp. 103–113). Cambridge, MA: MIT Press.

Marr, D. (1982). *Vision*. Cambridge, MA: MIT Press.

Martinez-Martinique, F. (2014). Systematicity and conceptual pluralism. In P. Calvo & J. Symons (Eds.), *The Architecture of Cognition* (pp. 305–333). Cambridge, MA: MIT Press.

Maynard Smith, J. (1978). Optimization theory in evolution. *Annual Review of Ecology and Systematics, 9*, 31–56.

Mayo, D. (1996). *Error and the Growth of Experimental Knowledge*. Chicago: University of Chicago Press.

Mayr, E. (1974). Behavior programs and evolutionary strategies: Natural selection sometimes favors a genetically "closed" behavior program, sometimes an "open" one. *American Scientist, 62*(6), 650–659.

McEchron, M. D., Bouwmeester, H., Tseng, W., Weiss, C., & Disterhoft, J. F. (1998). Hippocampectomy disrupts auditory trace fear conditioning and contextual fear conditioning in the rat. *Hippocampus, 8*(6), 638–646.

McFarland, D. (1996). Animals as cost-based robots. In M. Boden (Ed.), *The Philosophy of Artificial Life* (pp. 179–207). Oxford: Oxford University Press.

McGivern, R. F., Andersen, J., Byrd, D., Mutter, K. L., & Reilly, J. (2002). Cognitive efficiency on a match to sample task decreases at the onset of puberty in children. *Brain and Cognition, 50*(1), 73–89.

McLaughlin, B. P. (2009). Systematicity redux. *Synthese, 170*, 251–274.

Mele, A. (1987). Are intentions self-referential? *Philosophical Studies, 52*(3), 309–329.

Menary, R. (Ed.). (2010). *The Extended Mind*. Cambridge, MA: MIT Press.

Miller, M. B., & Bassler, B. L. (2001). Quorum sensing in bacteria. *Annual Review of Microbiology, 55*(1), 165–199.

Millikan, R. (1984). *Language, Thought, and Other Biological Categories.* Cambridge, MA: MIT Press.

Millikan, R. (1989). Biosemantics. *Journal of Philosophy, 86,* 281–297.

Millikan, R. (1990). Truth rules, hoverflies, and the Kripke-Wittgenstein paradox. *Philosophical Review, 99*(3), 323–353.

Millikan, R. (1996). On swampkinds. *Mind & Language, 11*(1), 70–130.

Millikan, R. (2002). *Varieties of Meaning.* Cambridge, MA: MIT Press.

Misyak, J. B., & Chater, N. (2014). Virtual bargaining: A theory of social decision-making. *Philosophical Transactions of the Royal Society B, 369,* 20130487.

Misyak, J. B., Melkonyan, T., Zeitoun, H., & Chater, N. (2014). Unwritten rules: Virtual bargaining underpins social interaction, culture, and society. *Trends in Cognitive Sciences, 18*(10), 512–519.

Mitani, J. C., & Watts, D. P. (2001). Why do chimpanzees hunt and share meat? *Animal Behaviour, 61*(5), 915–924.

Mithen, S. (1990). *Thoughtful Foragers: A Study of Prehistoric Decision Makers.* Cambridge: Cambridge University Press.

Moll, J., Zahn, R., de Oliveira-Souza, R., Krueger, F., & Grafman, J. (2005). The neural basis of human moral cognition. *Nature Reviews. Neuroscience, 6,* 799–809.

Montague, P. R., Dayan, P., Person, C., & Sejnowski, T. J. (1995). Bee foraging in uncertain environments using predictive hebbian learning. *Nature, 377*(6551), 725–728.

Morgan, A. (2014). Representations gone mental. *Synthese, 191,* 213–244.

Morgan, M. (2012). *The World in the Model.* Cambridge: Cambridge University Press.

Morillo, C. (1990). The reward event and motivation. *Journal of Philosophy, 87,* 169–186.

Nagel, T. (1970). *The Possibility of Altruism.* Princeton: Princeton University Press.

Neander, K. (1996). Swampman meets swampcow. *Mind & Language, 11*(1), 70–130.

Neander, K. (2006). Content for cognitive science. In G. F. Macdonald & D. Papineau (Eds.), *Teleosemantics* (pp. 167–194). Oxford: Oxford University Press.

Neubauer, A. C., & Fink, A. (2009). Intelligence and neural efficiency. *Neuroscience and Biobehavioral Reviews, 33*(7), 1004–1023.

Neubauer, A. C., Grabner, R. H., Fink, A., & Neuper, C. (2005). Intelligence and neural efficiency: Further evidence of the influence of task content and sex on the brain–IQ relationship. *Brain Research. Cognitive Brain Research, 25*(1), 217–225.

Newell, A., Shaw, J. C., & Simon, H. A. (1958). Elements of a theory of human problem solving. *Psychological Review, 65*(3), 151–166.

Nichols, S., & Stich, S. P. (2003). *Mindreading. An Integrated Account of Pretence, Self-Awareness, and Understanding Other Minds* (Vol. 114). Oxford University Press.

Nilsson, D., & Pelger, S. (1994). A pessimistic estimate of the time required for an eye to evolve. *Proceedings. Biological Sciences, 256*(1345), 53–55.

Noe, R., & Voelkl, B. (2013). Cooperation and biological markets: The power of partner choice. In K. Sterelny, R. Joyce, B. Calcott, & B. Fraser (Eds.), *Cooperation and Its Evolution* (pp. 131–152). Cambridge, MA: MIT Press.

O'Doherty, J., Dayan, P., Schultz, J., Deichmann, R., Friston, K., & Dolan, R. J. (2004). Dissociable roles of ventral and dorsal striatum in instrumental conditioning. *Science, 304*(5669), 452–454.

O'Reilly, R. C., Petrov, A. A., Cohen, J. D., Lebiere, C., Herd, S. A., & Kriete, T. (2014). How limited systematicity emerges: A computational cognitive neuroscience approach. In P. Calvo & J. Symons (Eds.), *The Architecture of Cognition* (pp. 191–225). Cambridge, MA: MIT Press.

Odling-Smee, F. J., Laland, K. N., & Feldman, M. W. (2003). *Niche Construction: The Neglected Process in Evolution.* Princeton: Princeton University Press.

Okasha, S. (2006). *Evolution and the Levels of Selection.* Oxford: Oxford University Press.

Orzack, S. H., & Sober, E. (1994). Optimality models and the test of adaptationism. *American Naturalist, 143*(3), 361–380.

Papineau, D. (1987). *Reality and Representation.* Oxford: Blackwell.

Papineau, D. (2003). *The Roots of Reason.* Oxford: Oxford University Press.

Pearce, J. (1994). Discrimination and categorization. In N. J. Mackintosh (Ed.), *Animal Learning and Conditioning* (pp. 109–134). San Diego: Academic Press.

Pence, C., & Ramsey, G. (2013). A new foundation for the propensity definition of fitness. *British Journal for the Philosophy of Science, 64*, 851–881.

Penn, D. C., Holyoak, K. J., & Povinelli, D. J. (2008). Darwin's mistake: Explaining the discontinuity between human and nonhuman minds. *Behavioral and Brain Sciences, 31*(2), 109–178.

Phelps, S. M., & Ophir, A. G. (2009). Monogamous brains and alternative tactics: Neuronal V1aR, space use, and sexual infidelity among male prairie voles. In R. Dukas

& J. M. Ratcliffe (Eds.), *Cognitive Ecology II* (pp. 156–176). Chicago: University of Chicago Press.

Piccinini, G. (2015). *Physical Computation: A Mechanistic Account*. Oxford: Oxford University Press.

Piccinini, G., & Bahar, S. (2013). Neural computation and the computational theory of cognition. *Cognitive Science, 34*, 453–488.

Piccinini, G., & Garson, J. (2014). Functions must be performed at appropriate rates in appropriate situations. *British Journal for the Philosophy of Science, 65*(1), 1–20.

Piccinini, G., & Scarantino, A. (2010). Computation vs. information processing: why their difference matters to cognitive science. *Studies in History and Philosophy of Science, 41*(3), 237–246.

Piccinini, G., & Scarantino, A. (2011). Information processing, computation, and cognition. *Journal of Biological Physics, 37*(1), 1–38.

Piccinini, G., & Schulz, A. (forthcoming). The Ways of Altruism.

Pinker, S. (1997). *How the Mind Works*. New York: Norton.

Pinker, S., & Bloom, P. (1990). Natural language and natural selection. *Behavioral and Brain Sciences, 13*(4), 707–784.

Platt, M. L., & Glimcher, P. W. (1999). Neural correlates of decision Variables in parietal cortex. *Nature, 400*, 233–238.

Polis, G. A., & Myers, C. A. (1985). A survey of intraspecific predation among reptiles and amphibians. *Journal of Herpetology, 19*(1), 99–107.

Pollock, J. (2006). *Thinking about Acting: Logical foundations for Rational Decision Making*. New York: Oxford University Press.

Povinelli, D. (2003). *Folk Physics for Apes: The Chimpanzee's Theory of How the World Works*. Oxford: Oxford University Press.

Prinz, J. (2002). *Furnishing the Mind: Concepts and their Perceptual Basis*. Cambridge, MA: MIT Press.

Prinz, J. (2006). Is the mind really modular? In R. Stainton (Ed.), *Contemporary Debates in Cognitive Science* (pp. 22–36). Oxford: Blackwell.

Pylyshyn, Z. (1999). Is vision continuous with cognition? The case for cognitive impenetrability of visual perception. *Behavioral and Brain Sciences, 22*, 341–423.

Ramnerö, J., & Törneke, N. (2015). On having a goal: Goals as representations or behavior. *Psychological Record, 65*(1), 89–99.

Ramsey, W. (2007). *Representation Reconsidered*. Cambridge: Cambridge University Press.

Ramsey, W. (2014). Systematicity and architectural pluralism. In P. Calvo & J. Symons (Eds.), *The Architecture of Cognition* (pp. 253–276). Cambridge, MA: MIT Press.

Ramsey, W. (forthcoming). Must cognition be representational? *Synthese.*

Rand, D. G. (2016). Cooperation, fast and slow: Meta-analytic evidence for a theory of social heuristics and self-interested deliberation. *Psychological Science, 27*(9), 1102–1206.

Rathbun, G. B., & Redford, K. (1981). Pedal scent-marking in the rufous elephant-shrew, Elephantulus rufescens. *Journal of Mammalogy, 62*(3), 635–637.

Reader, S. M., & Laland, K. N. (2003). Animal innovation: An introduction. In S. M. Reader & K. N. Laland (Eds.), *Animal Innovation* (pp. 3–37). Oxford: Oxford University Press.

Reed, E. S. (1996). *Encountering the World: Toward an Ecological Psychology.* Oxford: Oxford University Press.

Reeve, H. K., & Sherman, P. W. (1993). Adaptation and the goals of evolutionary research. *Quarterly Review of Biology, 68*(1), 1–32.

Reiss, J., & Frigg, R. (2009). The philosophy of simulations: Hot new issues or same old stew? *Synthese, 169,* 593–613.

Richardson, R. (2007). *Evolutionary Psychology as Maladapted Psychology.* Cambridge, MA: MIT Press.

Richerson, P., & Boyd, R. (2005). *Not By Genes Alone.* Chicago: University of Chicago Press.

Rieskamp, J., Busemeyer, J. R., & Mellers, B. A. (2006). Extending the bounds of rationality: Evidence and theories of preferential choice. *Journal of Economic Literature, 44,* 631–661.

Ristau, C. A. (1996). Aspects of the cognitive ethology of an injury-feigning bird, the piping plover. In D. Jamieson & M. Bekhoff (Eds.), *Readings in Animal Cognition* (pp. 79–89). Cambridge, MA: MIT Press.

Robson, A. J. (2001). Why would nature give individuals utility functions? *Journal of Political Economy, 109*(4), 900–914.

Robson, A. J., & Samuelson, L. (2008). The evolutionary foundations of preferences. In J. Benhabib, A. Bisin, & M. Jackson (Eds.), *The Handbook of Social Economics* (pp. 221–310). Amsterdam: Elsevier Press.

Rosenberg, K., & Trevathan, W. (2002). Birth, obstetrics and human evolution. *BJOG, 109*(11), 1199–1206.

Rosenzweig, M. L. (1996). Do animals choose habitats? In M. Bekhoff & D. Jamieson (Eds.), *Readings in Animal Cognition* (pp. 185–199). Cambridge, MA: MIT Press.

Roth-Nebelsick, A. (2001). Evolution and function of leaf venation architecture: A review. *Annals of Botany, 87*(5), 553–566.

Rowlands, M. (1999). *The Body in Mind.* Cambridge: Cambridge University Press.

Rowlands, M. (2010). *The New Science of the Mind: From Extended Mind to Embodied Phenomenology.* Cambridge, MA: MIT Press.

Royall, R. (1997). *Statistical Evidence—A Likelihood Paradigm.* Boca Raton, FL: Chapman and Hall.

Rumelhart, D., McClelland, J., & Group, P. R. (Eds.). (1986). *Parallel Distributed Processing* (Vol. I). Cambridge, MA: MIT Press.

Rupert, R. D. (1999). The best test theory of extension: First principle(s). *Mind & Language, 14,* 321–355.

Rupert, R. D. (2004). Challenges to the hypothesis of extended cognition. *Journal of Philosophy, 101*(8), 1–40.

Rupert, R. D. (2009). *Cognitive Systems and the Extended Mind.* Oxford: Oxford University Press.

Russon, A. E. (2003). Innovation and creativity in forest-living rehabilitant orangutans. In S. M. Reader & K. N. Laland (Eds.), *Animal Innovation* (pp. 279–306). New York: Oxford University Press.

Russon, A. E., & Andrews, K. (2010). Orangutan pantomime: Elaborating the message. *Biology Letters.* doi:10.1098/rsbl.2010.0564.

Rypma, B., Berger, J. S., Prabhakaran, V., Martin Bly, B., Kimberg, D. Y., Biswal, B. B., et al. (2006). Neural correlates of cognitive efficiency. *NeuroImage, 33*(3), 969–979.

Sachs, J., Mueller, U., Wilcox, T., & Bull, J. (2004). The evolution of cooperation. *Quarterly Review of Biology, 79*(2), 135–160.

Sack, L., & Scoffoni, C. (2013). Leaf venation: Structure, function, development, evolution, ecology and applications in the past, present and future. *New Phytologist, 198*(4), 983–1000.

Sadrieh, A., Guth, W., Hammerstein, P., Harnad, S., Hoffrage, U., Kuon, B., Munier, B. R., Todd, P. M., Warglien, M., & Weber, M. (2001). Group report: Is there evidence for an adaptive toolbox? In G. Gigerenzer & R. Selten (Eds.), *Bounded Rationality: The Adaptive Toolbox* (pp. 83–102). Cambridge, MA: MIT Press.

Samuels, R. (2006). Is the mind massively modular? In R. Stainton (Ed.), *Contemporary Debates in Cognitive Science* (pp. 37–56). Oxford: Blackwell.

Samuelson, L. (2001). Analogies, adaptation, and anomalies. *Journal of Economic Theory, 97,* 320–366.

Santos, E., & Noggle, C. A. (2011). Synaptic pruning. In S. Goldstein & J. A. Naglieri (Eds.), *Encyclopedia of Child Behavior and Development* (pp. 1464–1465). Boston, MA: Springer.

Santos, L. R., & Rosati, A. G. (2015). The evolutionary roots of human decision making. *Annual Review of Psychology, 66*(1), 321–347.

Savage, L. (1954). *The Foundations of Statistics.* New York: Dover.

Schroeder, T. (2004). *Three Faces of Desire.* Oxford: Oxford University Press.

Schroeder, T., Roskies, A. L., & Nichols, S. (2010). Moral motivation. In J. Doris (Ed.), *Moral Psychology Handbook* (pp. 72–110). Oxford University Press.

Schultz, W., Tremblay, L., & Hollerman, J. (2000). Reward processing in primate orbitofrontal cortex and basal ganglia. *Cerebral Cortex, 10,* 272–283.

Schulz, A. (2008). Structural flaws: Massive modularity and the argument from design. *British Journal for the Philosophy of Science, 59,* 733–743.

Schulz, A. (2010). It takes two: Sexual strategies and game theory. *Studies in History and Philosophy of Science Part C: Studies in History and Philosophy of Biological and Biomedical Sciences, 41*(1), 41–49.

Schulz, A. (2011a). The adaptive importance of cognitive efficiency: An alternative theory of why we have beliefs and desires. *Biology & Philosophy, 26*(1), 31–50.

Schulz, A. (2011b). Gigerenzer's evolutionary arguments against rational choice theory: An assessment. *Philosophy of Science, 78*(5), 1272–1282.

Schulz, A. (2011c). Sober & Wilson's evolutionary arguments for psychological altruism: A reassessment. *Biology & Philosophy, 26,* 251–260.

Schulz, A. (2012). Heuristic evolutionary psychology. In K. Plaisance & T. Reydon (Eds.), *Philosophy of Behavioral Biology* (pp. 217–234). Dordrecht: Springer.

Schulz, A. (2013a). The benefits of rule following: A new account of the evolution of desires. *Studies in History and Philosophy of Science Part C: Studies in History and Philosophy of Biological and Biomedical Sciences, 44*(4, Part A), 595–603.

Schulz, A. (2013b). Exaptation, adaptation, and evolutionary psychology. *History and Philosophy of the Life Sciences, 35,* 193–212.

Schulz, A. (2013c). Overextension: The extended mind and arguments from evolutionary biology. *European Journal for Philosophy of Science, 3*(2), 241–255.

Schulz, A. (2013d). Selection, drift, and independent contrasts: Defending the methodological foundations of the FIC. *Biological Theory, 7*(1), 38–47.

Schulz, A. (2015a). The heuristic defense of scientific models: An incentive-based assessment. *Perspectives on Science, 23,* 424–442.

Schulz, A. (2015b). Preferences vs. desires: Debating the fundamental structure of conative states. *Economics and Philosophy*, *31*(02), 239–257.

Schulz, A. (2016a). Altruism, egoism, or neither: A cognitive-efficiency-based evolutionary biological perspective on helping behavior. *Studies in History and Philosophy of Science Part C: Studies in History and Philosophy of Biological and Biomedical Sciences*, *56*, 15–23.

Schulz, A. (2016b). Firms, agency, and evolution. *Journal of Economic Methodology*, *23*(1), 57–76.

Scott, G. (2005). *Essential Animal Behavior*. Oxford: Blackwell.

Seeley, T. D., Visscher, P. K., & Passino, K. M. (2006). Group decision making in honey bee swarms. *American Scientist*, *94*(3), 220–229.

Selfridge, O., & Neisser, U. (1960). Pattern recognition by machine. *Scientific American*, *203*, 60–68.

Sellars, W. (1963). *Science, Perception, and Reality*. London: Routledge.

Seyfarth, R., & Cheney, D. (1992). Inside the mind of a monkey. *New Scientist* (January 4), 25–29.

Shackelford, T. K., Schmitt, D. P., & Buss, D. M. (2005). Universal dimensions of human mate preferences. *Personality and Individual Differences*, *39*(2), 447–458.

Shafer, G. (1976). *A Mathematical Theory of Evidence*. Princeton: Princeton University Press.

Shapiro, L. (1999). Presence of Mind. In V. Hardcastle (Ed.), *Biology Meets Psychology: Constraints, Connections, Conjectures* (pp. 83–98). Cambridge, MA: MIT Press.

Shapiro, L. (2004). *The Mind Incarnate*. Cambridge, MA: MIT Press.

Shapiro, L. (2011a). *Embodied Cognition*. London: Routledge.

Shapiro, L. (2011b). James Bond and the barking dog: Evolution and extended cognition. *Philosophy of Science*, *77*, 410–418.

Shea, N. (2014). Reward prediction error signals are meta-representational. *Noûs*, *48*(2), 314–341.

Shepard, R. N. (1981). Psychological relations and psychophysical scales: On the status of "direct" psychophysical measurement. *Journal of Mathematical Psychology*, *24*(1), 21–57.

Sherman, P. W., Reeve, H. K., & Pfennig, D. W. (1997). Recognition systems. In J. Krebs & N. Davies (Eds.), *Behavioural Ecology: An Evolutionary Perspective* (pp. 69–96). Oxford: Blackwell.

Shettleworth, S. J. (2002). Spatial behavior, food storing, and the modular mind. In M. Bekhoff, C. Allen, & G. Burghardt (Eds.), *The Cognitive Animal* (pp. 123–128). Cambridge: MIT Press.

Shettleworth, S. J. (2009). *Cognition, Evolution, and Behavior.* Oxford: Oxford University Press.

Silberstein, M., & Chemero, A. (2012). Complexity and extended phenomenological-cognitive systems. *Topics in Cognitive Science, 4*(1), 35–50.

Simmons, S. L., Bazylinski, D. A., & Edwards, K. J. (2006). South-seeking magnetotactic bacteria in the Northern hemisphere. *Science, 311*(5759), 371–374.

Simon, H. A. (1957). *Models of Bounded Rationality.* Cambridge, MA: MIT Press.

Simon, H. A. (1962). The architecture of complexity. *Proceedings of the American Philosophical Society, 106*(6), 467–482.

Sinhababu, N. (2013). Distinguishing belief and imagination. *Pacific Philosophical Quarterly, 94*, 152–165.

Skerry, A. E., Carey, S. E., & Spelke, E. S. (2013). First-person action experience reveals sensitivity to action efficiency in prereaching infants. *Proceedings of the National Academy of Sciences of the United States of America, 110*(46), 18728–18733.

Skyrms, B. (1996). *Evolution and the Social Contract.* Cambridge: Cambridge University Press.

Skyrms, B. (2004). *The Stag Hunt and the Evolution of Social Structure.* Cambridge: Cambridge University Press.

Skyrms, B. (2010). *Signals: Evolution, Learning, and Information.* Oxford: Oxford University Press.

Sloman, S. A. (1996). The empirical case for two systems of reasoning. *Psychological Bulletin, 119*, 3–22.

Smith, L. B., & Thelen, E. (1994). *A Dynamic Systems Approach to the Development of Cognition and Action.* Cambridge, MA: MIT Press.

Smith, M. (1987). The Humean theory of motivation. *Mind, 96*(381), 36–61.

Smith, W. J. (1990). Communication and expectations: A social process and the cognitive operations it depends upon and influences. In M. Bekoff & D. Jamieson (Eds.), *Interpretation and Explanation in the Study of Animal Behavior* (Vol. I: Interpretation, Intentionality, and Communication, pp. 234–253). Boulder, CO: Westview Press.

Smuts, B., & Smuts, R. (1993). Male aggression and sexual coercion of females in non-human primates and other mammals: Evidence and Theoretical Implications. In

P .J. B. Slater, J. S. Rosenblatt, C. T. Snowdon, & M. Milinski (Eds.), *Advances in the Study of Behavior* (Vol. 22, pp. 1–63).

Sober, E. (1981). The evolution of rationality. *Synthese, 46,* 95–120.

Sober, E. (1984). *The Nature of Selection.* Cambridge: Cambridge University Press.

Sober, E. (1994a). The adaptive advantage of learning and a priori prejudice. *Ethology and Sociobiology, 15*(1), 55–56.

Sober, E. (1994b). *The Primacy of Truth Telling and the Evolution of Lying From a Biological Point of View* (pp. 71–92). Cambridge: Cambridge University Press.

Sober, E. (1997). Is the mind an adaptation for coping with environmental complexity? *Biology & Philosophy, 12,* 539–550.

Sober, E. (1998a). Black box inference. *British Journal for the Philosophy of Science, 49,* 469–498.

Sober, E. (1998b). Morgan's canon. In D. D. C. C. Allen (Ed.), *The Evolution of Mind* (pp. 224–242). New York: Oxford University Press.

Sober, E. (2000). *Philosophy of Biology* (2nd ed.). Boulder, CO: Westview Press.

Sober, E. (2001). The Two Faces of Fitness. In R. Singh, D. Paul, C. Krimbas, & J. Beatty (Eds.), *Thinking about Evolution: Historical, Philosophical, and Political Perspectives* (pp. 309–321). Cambridge: Cambridge University Press.

Sober, E. (2008). *Evidence and Evolution.* Cambridge: Cambridge University Press.

Sober, E. (2010). Natural selection, causality, and laws: What Fodor and Piatelli-Palmarini got wrong. *Philosophy of Science, 77,* 594–607.

Sober, E., & Wilson, D. S. (1998). *Unto Others: The Evolution and Psychology of Unselfish Behavior.* Cambridge, MA: Harvard University Press.

Sopher, B., & Gigliotti, G. (1993). Intransitive cycles: Rational choice or random error? *Theory and Decision, 35,* 311–336.

Sperber, D. (2005). Modularity and relevance: How can a massively modular mind be flexible and context-sensitive. In P. Carruthers, S. Laurence, & S. Stich (Eds.), *The Innate Mind* (pp. 53–68). Oxford: Oxford University Press.

Spohn, W. (2012). *The Laws of Belief.* Oxford: Oxford University Press.

Sporns, O., Tononi, G., & Edelman, G. M. (2000). Connectivity and complexity: The relationship between neuroanatomy and brain dynamics. *Neural Networks, 13*(8–9), 909–922.

Sprevak, M. (2009). Extended cognition and functionalism. *Journal of Philosophy, 106,* 503–527.

Spurrett, D. (2015). The natural history of desire. *South African Journal of Philosophy*, *34*, 304–313.

Stampe, D. (1986). Verification and a causal account of meaning. *Synthese, 69*, 107–137.

Stanford, C. B. (1995). Chimpanzee hunting behavior and human Evolution. *American Scientist, 83*(3), 256–261.

Stanovich, K. E. (2004). *The Robot's Rebellion: Finding Meaning in the Age of Darwin.* Chicago: University of Chicago Press.

Stanovich, K. E., & West, R. F. (2000). Individual differences in reasoning: implications for the rationality debate. *Behavioral and Brain Sciences, 23*, 645–665.

Stegmann, U. (2009). A consumer-based teleosemantics for animal signals. *Philosophy of Science, 76*, 864–875.

Stephens, C. L. (2001). When is it selectively advantageous to have true beliefs: Sandwiching the better safe than sorry argument. *Philosophical Studies, 105*, 161–189.

Stephens, D. W. (1989). Variance and the value of information.pdf. *American Naturalist, 134*(1), 128–140.

Sterelny, K. (1999). Situated agency and the descent of desire. In V. Hardcastle (Ed.), *Where Biology Meets Psychology: Constraints, Connections, Conjectures* (pp. 203–220). Cambridge, MA: MIT Press.

Sterelny, K. (2001). *The Evolution of Agency and Other Essays.* Cambridge, MA: Cambridge University Press.

Sterelny, K. (2003). *Thought in a Hostile World: The Evolution of Human Cognition.* Oxford: Wiley-Blackwell.

Sterelny, K. (2012). *The Evolved Apprentice: How Evolution Made Humans Unique.* Cambridge, MA: MIT Press.

Sterelny, K., & Griffiths, P. (1999). *Sex and Death.* Chicago: University of Chicago Press.

Stich, S. (1983). *From Folk Psychology to Cognitive Science.* Cambridge, MA: The MIT Press.

Stich, S. (1990). *The Fragmentation of Reason.* Cambridge, MA: MIT Press.

Stich, S. (2007). Evolution, altruism and cognitive architecture: A critique of Sober and Wilson's argument for psychological altruism. *Biology & Philosophy, 22*, 267–281.

Stich, S., Doris, J., & Roedder, E. (2010). Altruism. In J. M. Doris & The Moral Psychology Research Group (Eds.), *The Moral Psychology Handbook* (pp. 147–205). Oxford: Oxford University Press.

Stotz, K. (2010). Human nature and cognitive-developmental niche construction. *Phenomenology and the Cognitive Sciences, 9*(4), 483–501.

Strassmann, J. E., Gilbert, O. M., & Queller, D. C. (2011). Kin discrimination and cooperation in microbes. *Annual Review of Microbiology, 65*, 349–367.

Theraulaz, G., & Bonabeau, E. (1999). A brief history of stigmergy. *Artificial Life, 5*(2), 97–116.

Thometz, N. M., Tinker, M. T., Staedler, M. M., Mayer, K. A., & Williams, T. M. (2014). Energetic demands of immature sea otters from birth to weaning: Implications for maternal costs, reproductive behavior and population-level trends. *Journal of Experimental Biology, 217*(Pt 12), 2053–2061.

Thompson, E. (2007). *Mind in Life: Biology, Phenomenology, and the Sciences of Mind.* Cambridge, MA: Harvard University Press.

Todd, P. (2001). Fast and frugal heuristics for environmentally bounded minds. In G. Gigerenzer & R. Selten (Eds.), *Bounded Rationality: The Adaptive Toolbox* (pp. 51–70). Cambridge, MA: MIT Press.

Tooby, J., & Cosmides, L. (1992). The psychological foundations of culture. In J. Barkow, L. Cosmides, & J. Tooby (Eds.), *The Adapted Mind* (pp. 19–136). Oxford: Oxford University Press.

Tooby, J., Cosmides, L., & Barrett, H. C. (2005). Resolving the debate on innate ideas: Learnability constraints and the evolved interpenetration of motivational and conceptual functions. In P. Carruthers, S. Laurence, & S. Stich (Eds.), *The Innate Mind: Structure and Contents* (pp. 305–337). Oxford: Oxford University Press.

Trimmer, P. C., Houston, A. I., Marshall, J. A., Bogacz, R., Paul, E. S., Mendl, M. T., et al. (2008). Mammalian choices: Combining fast-but-inaccurate and slow-but-accurate decision-making systems. *Proceedings. Biological Sciences, 275*(1649), 2353–2361.

Trimmer, P. C., Houston, A. I., Marshall, J. A., Mendl, M. T., Paul, E. S., & McNamara, J. M. (2011). Decision-making under uncertainty: Biases and Bayesians. *Animal Cognition, 14*(4), 465–476.

Trivers, R. (1971). The evolution of reciprocal altruism. *Quarterly Review of Biology, 46*, 35–57.

Trivers, R. (1974). Parent-offspring conflict. *American Zoologist, 14*, 247–262.

Trivers, R., & Willard, D. E. (1973). Natural selection of parental ability to vary the sex ratio of offspring. *Science, 179*(4068), 90–92.

Tsai, R.-C., & Bockenholt, U. (2006). Modeling intransitive preferences: A random-effects approach. *Journal of Mathematical Psychology, 50*, 1–14.

Tulving, E. (1985). How many memory systems are there? *American Psychologist, 40,* 385–398.

Tversky, A., & Kahneman, D. (1974). Judgment under uncertainty: Heuristics and biases. *Science, 185,* 1124–1131.

Van Gelder, T. (1995). What might cognition be, if not computation? *Journal of Philosophy, 91,* 345–381.

Van Gils, J. A., Schenk, I. W., Bos, O., & Piersma, T. (2003). Incompletely informed shorebirds that face a digestive constraint maximize net energy gain when exploiting patches. *American Naturalist, 161*(5), 777–793.

Varela, F. J., Thompson, E., & Rosch, E. (1991). *The Embodied Mind: Cognitive Science and Human Experience.* Cambridge, MA: MIT Press.

Waite, T. (2001). Intransitive preferences in hoarding gray jays (Perisoreus canadensis). *Behavioral Ecology and Sociobiology, 50,* 116–121.

Wake, D. B. (1991). Homoplasy: The result of natural selection, or evidence of design limitations. *American Naturalist, 138*(3), 543–567.

Wall, D. (2009). Are there passive desires? *Dialectica, 63,* 133–155.

Walsh, D. (1997). Review of "Complexity and the Function of Mind in Nature" by Peter Godfrey-Smith. *British Journal for the Philosophy of Science, 48*(4), 613–617.

Walter, S. (2009). Review: Robert C. Richardson: "Evolutionary Psychology as Maladapted Psychology". *Mind, 118*(470), 523–527.

Weisberg, M. (2007). Who is a modeller? *British Journal for the Philosophy of Science, 58,* 207–233.

Weisberg, M. (2007a). Three kinds of idealization. *Journal of Philosophy, 104,* 639–659.

Weisberg, M. (2013). *Simulation and Similarity: Using Models to Understand the World.* Oxford: Oxford University Press.

West, S. A., Griffin, A. S., & Gardner, A. (2007). Social semantics: Altruism, cooperation, mutualism, strong reciprocity and group selection. *Journal of Evolutionary Biology, 20*(2), 415–432.

West, S. A., Griffin, A. S., & Gardner, A. (2008). Social semantics: How useful has group selection been? *Journal of Evolutionary Biology, 21*(1), 374–385.

West-Eberhard, M. J. (1992). Adaptation: Current usages. In E. Fox Keller & E. Lloyd (Eds.), *Keywords in Evolutionary Biology* (pp. 13–18). Cambridge, MA: Harvard University Press.

Whiten, A. (1995). When does smart behavior-reading become mind-reading? In P. Carruthers & P. Smith (Eds.), *Theories of Theories of Mind* (pp. 277–292). Cambridge: Cambridge University Press.

Whiten, A. (2013). Humans are not alone in computing how others see the world. *Animal Behaviour, 86*(2), 213–221.

Whiten, A., & Byrne, R. W. (Eds.). (1997). *Machiavellian Intelligence II: Extensions and Evaluations*. Cambridge: Cambridge University Press.

Wilder, H. (1996). Interpretive cognitive ethology. In M. Bekhoff & D. Jamieson (Eds.), *Readings in animal cognition* (pp. 29–45). Cambridge, MA: MIT Press.

Williams, A. M., & Hodges, N. J. (2004). *Skill Acquisition in Sport: Research, Theory and Practice*. Taylor & Francis.

Williams, G. C. (1966). *Adaptation and Natural Selection*. Princeton, NJ: Princeton University Press.

Wilson, E. O. (1971). *The Insect Societies*. Cambridge, MA: Harvard University Press.

Wilson, R. (2004). *Boundaries of the Mind*. Cambridge: Cambridge University Press.

Wilson, R. (2010). Meaning making and the mind of the externalist. In R. Menary (Ed.), *The Extended Mind* (pp. 167–188). Cambridge: MIT Press.

Wilson, T. D. (2002). *Strangers to Ourselves*. Cambridge, MA: Belknap.

Xu, X., Zhou, Z., Wang, X., Kuang, X., Zhang, F., & Du, X. (2003). Four-winged dinosaurs from China. *Nature, 421*(6921), 335–340.

Yager, R. R. (1987). On the Dempster-Shafer framework and new combination rules. *Information Sciences, 41*(2), 93–137.

Zaitchik, D., Iqbal, Y., & Carey, S. (2014). The effect of executive function on biological reasoning in young children: An individual differences study. *Child Development, 85*(1), 160–175.

Zangwill, N. (2008). Besires and the motivation debate. *Theoria, 74*, 50–59.

Index